AF597479

INTERNATIONAL CENTRE FOR MECHANICAL SCIENCES

COURSES AND LECTURES - No. 324

ADAPTIVE SIGNAL PROCESSING

EDITED BY

L.D. DAVISSON
UNIVERSITY OF MARYLAND

G. LONGO
UNIVERSITY OF TRIESTE

SPRINGER - VERLAG

WIEN - NEW YORK

Le spese di stampa di questo volume sono in parte coperte da
contributi del Consiglio Nazionale delle Ricerche.

This volume contains 14 illustrations.

Printed in Italy

In order to make this volume available as economically and as rapidly as possible the authors' typescripts have been reproduced in their original forms. This method unfortunately has its typographical limitations but it is hoped that they in no way distract the reader.

ISBN 3-211-82333-6 Springer Verlag Wien-New York
ISBN 0-387-82333-6 Springer Verlag New York-Wien

PREFACE

This volume contains most of the contributions to the advanced school on adaptive prediction methods held at the International Centre for Mechanical Sciences, Udine, Italy, in July 1990.

Thomas C. Butash and Lee D. Davisson address the issue of adaptive linear prediction, which is employed to solve a variety of problems, varying from adaptive source coding to autoregressive spectral estimation. An adaptive linear predictor is realized by a linear combination of a finite number M of observations preceding each sample to be predicted, where the coefficients defining the predictor are estimated on the basis of the preceding M+N observations, in an attempt to optimize the predictor's performance.

The performance of an adaptive linear predictor is formulated by Butash and Davisson in a series of theorems, with and without the Gaussian assumption, under the hypotheses that its coefficients are derived from either the (single) observation sequence to be predicted (dependent case) or a second, statistically independent realization (independent case).

Three recently developed general methodologies for designing signal predictors under nonclassical operating conditions are presented in the contribution by H. Vincent Poor, namely the robust prediction, the high-speed Levinson modeling and the approximate conditional mean (ACM) nonlinear prediction. The ACM nonlinear prediction is discussed in the specific context of interference suppression in wideband digital communication systems; and this approach is seen to be much more effective than the traditional fixed and adaptive linear prediction approaches usually applied to such problems.

Mati Wax presents the key concepts and techniques for detecting, localizing and beamforming multiple narrowband sources by passive sensor arrays. Adaptive processing in sensor arrays arises in a variety of fields, ranging from radar, sonar, oceanography and seismology to medical imaging and radio-astronomy. The sources considered are narrowband and arbitrarily correlated, and optimal and suboptimal solutions to the problems of detection, localization and beamforming are presented.

Special coding algorithms and techniques based on the use of linear prediction now permit high-quality voice reproduction at low bit rates. Allen Gersho reviews some of the main ideas underlying the most interesting algorithms. The concepts of removing redundancy by linear prediction, of linear predictive coding, adaptive prediction coding, and vector quantization are discussed and then the author proceeds to more specific illustrations of the linear prediction methods in the field of speech coding, as well as some more recent work on nonlinear prediction of speech and its potential for the future of speech coding.

L.D. Davisson

G. Longo

CONTENTS

ADAPTIVE LINEAR PREDICTION AND PROCESS ORDER IDENTIFICATION

T.C. Butash, L.D. Davisson
University of Maryland, College Park, MD, USA

ABSTRACT

Adaptive linear predictors are employed to provide solutions to problems ranging from adaptive source coding to autoregressive (AR) spectral estimation. In such applications, an adaptive linear predictor is realized by a linear combination of a finite number, M, of the observations immediately preceding each sample to be predicted, where the coefficients defining the predictor are "adapted to", or estimated on the basis of, the preceding $N+M$ observations in an attempt to continually optimize the predictor's performance. This performance is thus inevitably dictated by the predictor's order, M, and the length of its learning period, N.

We formulate the adaptive linear predictor's MSE performance in a series of theorems, with and without the Gaussian assumption, under the hypotheses that its coefficients are derived from either the (single) observation sequence to be predicted (*dependent case*), or a second, statistically independent realization (*independent case*). The established theory on adaptive linear predictor performance and order selection is reviewed, including the works of Davisson (Gaussian, dependent case), and Akaike (AR, independent case). Results predicated upon the independent case hypothesis (e.g., Akaike's FPE procedure) are shown to incur substantial error under the dependent case conditions prevalent in typical adaptive prediction environments. Similarly, theories based on the Gaussian assumption are found to suffer a loss in accuracy which is proportional to the deviation, of the probability law governing the process in question, from the normal distribution.

We develop a theory on the performance of, and an optimal order selection criterion for, an adaptive linear predictor which is applicable under the dependent case, in environments where the Gaussian assumption is not necessarily justifiable.

---

† Thomas C. Butash is also affiliated with the IBM Corporation, Manassas, Va. 22110.

Chapter 1
Introduction to and Statement of the Problem

1.0 Introduction

Adaptive linear prediction of discrete-time stochastic processes has found widespread application [1-6] in such diverse areas as speech and data compression, narrowband interference rejection and equalization filtering, forecasting, and autoregressive spectral estimation. Such applications share the common underlying objective of minimizing the mean square (prediction) error (MSE) incurred under the implementation constraint of a linear predictor structure. Under these criteria, the optimal, or Linear Minimum Mean Square Error (LMMSE), predictor is well known to be uniquely determined, via Wiener-Kolmogorov filtering theory, by the mean and covariance of the process on which the predictor must operate. Unfortunately, these moments are seldom known *a priori* in the subject, as well as most other, applications of practical interest.

It is in such applications that adaptive linear predictors, or autoregressive (AR) models, are employed. The adaptive LMMSE predictor is realized by a linear combination of a finite number, M, of the observations immediately preceding each sample to be predicted, where the weights or coefficients which define the predictor are determined or "adapted" on the basis of the preceding $N+M$ observations in such a manner as to continually seek to minimize the mean square prediction error.

Although several techniques for determining the adaptive LMMSE predictor coefficients have been proposed, the most commonly employed appears to be a simple least mean squares (LMS) estimate; i.e., the M coefficients are determined such that the sample mean square prediction error (mean square residual error) over the previous N observations is minimized. The adaptive LMMSE predictor's memory, or order, M, and the length of its learning period, N, are typically chosen and fixed *a priori* based on various measures of the desired mean square prediction error performance, the anticipated degree of nonstationarity of the process to be predicted, and limits, imposed by the application in question, on the computational complexity of the adaptation algorithm.

The performance of adaptive linear predictors, including the LMS adaptive LMMSE predictor, relative to the mean square error cost function, is in general critically dependent on the choice of M and N. Recognizing this, several authors [1, 7-39] have investigated the dependence of the adaptive linear predictor's asymptotic performance (i.e., as $N \rightarrow \infty$) on its order and learning period. Davisson [1,7,8], Akaike [9-14], Parzen [15-16], Schwartz [19], Rissanen [22], Hannan and Quinn [23-24], and Wax [35] carried these analyses to their logical conclusion by proposing techniques for selecting the predictor's order and learning period which are designed to optimize the estimated performance of the resulting predictor or AR model. The assumptions, analytical limitations, and conclusions of these studies provide a point of departure for the theory developed in Chapters 2 through 4.

The class of problems addressed by, and the objectives and approach of, this treatise are outlined in the sections which follow.

1.1 Problem Definition[1]

Given $\{S_k\}_{k=-\infty}^{+\infty}$ a real-valued, second order, zero mean[2], stochastic sequence, the linear minimum mean square error (LMMSE) predictor, of order M, for S_n on the basis of $\underline{S}_n = (S_{n-1}, \ldots, S_{n-M})^T$ is defined by

$$\hat{S}_{n_{LMMSE}} = \underline{\beta}_{M,\infty}^T(n)\, \underline{S}_n \tag{1.1}$$

where

$$\underline{\beta}_{M,\infty}(n) = arg \min_{\underline{\beta} \in R^M} E(S_n - \underline{\beta}^T \underline{S}_n)^2 .$$

It is well known[3] that the Orthogonality Principle provides the necessary and sufficient conditions for the LMMSE predictor in the form of the normal equations:

$$E(S_n - \underline{\beta}_{M,\infty}^T(n)\, \underline{S}_n)\underline{S}_n = \underline{0}$$

or equivalently,

$$R_M\, \underline{\beta}_{M,\infty}(n) = \underline{G}_M \tag{1.2}$$

where

$$R_M = E\underline{S}_n\, \underline{S}_n^T = \{\mathrm{cov}(S_{n-i}, S_{n-j});\ i, j = 1,\ldots,M\}$$

$$\underline{G}_M = ES_n\, \underline{S}_n = \{\mathrm{cov}(S_n, S_{n-j});\ j = 1,\ldots,M\} .$$

It immediately follows that the mean square prediction error of the Mth order LMMSE predictor is given by

$$\sigma_n^2[M,\infty] = E(S_n - \underline{\beta}_{M,\infty}^T(n)\underline{S}_n)^2 = ES_n^2 - \underline{\beta}_{M,\infty}^T(n)\underline{G}_M . \tag{1.3}$$

Thus the LMMSE predictor and its performance are completely characterized by the first- and second-order moments of the process on which the predictor is to operate.

However, in many applications these moments are neither known nor can they be estimated *a priori* with any degree of certainty. In such applications any realizable linear predictor, whose performance is to approach that of the LMMSE predictor, must be designed such that its coefficients are "learned from," or estimated on the basis of, a finite number of observations immediately preceding each point to be predicted. These adaptive LMMSE predictors thus continually adjust their impulse responses, according to estimates of the encountered statistics of the processes on which they operate, in seeking to minimize their mean square prediction error. (It is interesting to note that such predictors are, in actuality, inherently *nonlinear*, although by convention they are said to be linear.)

---

[1] We employ the notation of [7,8] throughout the sequel.

[2] Or known mean, in which case it is assumed here that the mean in question is removed from the process to be predicted.

[3] As a special case of Wiener-Kolmogorov filtering theory.

A common adaptation strategy entails employing the LMS estimate, $\underline{\beta}_{M,N}$, of the optimal LMMSE weights $\underline{\beta}_{M,\infty}$, based on the $N+M$ observations which precede each point to be predicted. Thus, $\underline{\beta}_{M,N}$ minimizes the sample mean square prediction error over the preceding N observations:

$$\underline{\beta}_{M,N}(n) = arg \min_{\underline{\beta} \in R^M} \frac{1}{N} \sum_{k=1}^{N} (S_{n-k} - \underline{\beta}^T \underline{S}_{n-k})^2 .$$

The LMS estimate $\underline{\beta}_{M,N}$ of the LMMSE coefficients is given by the solution to the Yule-Walker equation:

$$r_M \underline{\beta}_{M,N}(n) = \underline{q}_M \tag{1.4}$$

where

$$r_M = \frac{1}{N} \sum_{k=1}^{N} \underline{S}_{n-k} \underline{S}_{n-k}^T = \{ r(i,j) = \frac{1}{N} \sum_{k=1}^{N} S_{n-k-i} S_{n-k-j} \ ; \ i,j = 1,\ldots,M \}$$

$$\underline{q}_M = \frac{1}{N} \sum_{k=1}^{N} S_{n-k} \underline{S}_{n-k} = \{ r(0,j) = \frac{1}{N} \sum_{k=1}^{N} S_{n-k} S_{n-k-j} \ ; \ j = 1,\ldots,M \}$$

are the sample covariances of the process. Thus the sample mean square prediction (LMS residual or fitting) error becomes

$$F^2(M,N) = \frac{1}{N} \sum_{k=1}^{N} (S_{n-k} - \underline{\beta}_{M,N}^T \underline{S}_{n-k})^2 = \frac{1}{N} \sum_{k=1}^{N} S_{n-k}^2 - \underline{\beta}_{M,N}^T \underline{q}_M \tag{1.5}$$

and the LMS adaptive LMMSE predictor and its associated mean square prediction error are defined by

$$\hat{S}_n = \underline{\beta}_{M,N}^T(n) \underline{S}_n \tag{1.6}$$

$$\sigma_n^2[M,N] = E(S_n - \hat{S}_n)^2 . \tag{1.7}$$

Comparing (1.2) and (1.3) with (1.4) and (1.5) reveals that the LMS adaptive predictor and its LMS error may be obtained by substituting sample mean estimates (computed over the adaptive predictor's learning period) for the second order moments which, if known, would uniquely determine the optimal predictor and its MMSE in (1.2) and (1.3), respectively. Hence, the LMS adaptive predictor and an estimate of its mean square error are given by the solution to the same set of linear equations which define the optimal predictor and its MMSE, with sample mean estimates[4], of the unknown second order moments, employed to realize the required adaptive solution. This intuitively pleasing result is reinforced by the realization that the LMS adaptive predictor minimizes the sample mean square prediction error over the N observations which immediately precede each sample to be predicted. The inherent computational simplicity of the LMS adaptive predictor's linear structure, when combined with its intellectually appealing characteristics and interpretation, undoubtedly explains the widespread use of the LMS criterion in adaptive prediction applications.

---

[4] It is reassuring to recognize that the LMS estimates of the optimal predictor's coefficients and MMSE are asymptotically equivalent to Maximum Likelihood estimates if the process to be predicted is Mth order Gauss-Markov.

Acknowledging this almost universal acceptance, we dedicate the remainder of this treatise to the investigation of adaptive linear predictors determined under the LMS adaptation criterion. This analytical restriction does not, however, appreciably limit the applicability of the conclusions which follow, for Fuller and Hasza have shown [25] that asymptotic results obtained on the LMS adaptive predictor's convergence rate, and mean square error (1.7), apply to any adaptive predictor $\hat{\underline{\beta}}_{M,N}$ such that

$$E \parallel \underline{\beta}_{M,N} - \hat{\underline{\beta}}_{M,N} \parallel^4 = O(N^{-4}) .$$

In this context, the terms *LMS adaptive predictor* and *adaptive linear predictor* are employed interchangeably hereinafter.

The adaptive linear predictor's mean square prediction error, (1.7), can be expanded in terms of the difference between the LMS estimate, $\underline{\beta}_{M,N}$, and the optimal LMMSE weights, $\underline{\beta}_{M,\infty}$, as

$$\sigma_n^2[M,N] = \sigma_n^2[M,\infty] + 2E(S_n - \underline{\beta}_{M,\infty}^T(n)\,\underline{S}_n)\Delta\underline{\beta}_{M,N}^T\underline{S}_n + E(\Delta\underline{\beta}_{M,N}^T\underline{S}_n)^2 \quad (1.8)$$

where

$$\Delta\underline{\beta}_{M,N} = \underline{\beta}_{M,\infty} - \underline{\beta}_{M,N}$$

and the sum of the second two terms on the right-hand side of (1.8) is, from (1.1) and (1.3), nonnegative and reflects the performance degrading effect of adaptation errors incurred in estimating the optimal weights.

Throughout the sequel, we will assume that the process in question is not perfectly predictable[5], in the mean square error sense, with a linear predictor structure; i.e., it is assumed that for all $1 \leq M \leq \infty$, $\sigma^2[M,\infty] > 0$, which, in turn, implies that $R_M > 0$ for all $M < \infty$. It immediately follows that R_M is nonsingular and therefore admits an unique solution for the Mth order optimal predictor as given by $\underline{\beta}_{M,\infty} = R_M^{-1}\underline{G}_M$. Although these results are well known consequences of the subject assumption, we nevertheless provide their proof, for completeness, in Appendix 1A. The limitations imposed by the adoption of this assumption are discussed in section 4.1.1.

In view of the preceding developments, and recalling that our lack of knowledge of the processes second order moments necessitated an adaptive approach, four questions concerning the performance of an Mth order LMS adaptive predictor arise immediately:

1) Can the optimal and LMS adaptive predictors' mean square prediction errors be estimated reliably (at least in the asymptotic sense)?

2) Under what conditions and how rapidly does the LMS adaptive predictor, as defined by (1.4), (1.6), converge in the mean square to the optimal LMMSE predictor $\underline{\beta}_{M,\infty}$ (or, equivalently, under what conditions and how rapidly does the mean square error of the LMS adaptive predictor converge to its lower bound (1.3)), as the length of the predictor's learning period $N \rightarrow \infty$?

[5] Doob [40] defines such processes as *regular* or *nondeterministic*.

3) What design procedure should be employed to select the order, M, and the learning period, N, of the LMS adaptive predictor to complete its definition given
(i) an acceptable level of mean square prediction error performance
(ii) the anticipated degree of nonstationarity of the process to be predicted, and
(iii) implementation constraints on the computational complexity of the resulting predictor's LMS adaptation algorithm?

4) What minimal set of constraints or assumptions must be imposed on the class of processes to be predicted and on the implementation of the LMS adaptive predictor in order to facilitate the derivation of results, relative to issues 1) through 3) above, which prove useful in most, if not all, practical applications of interest?

These questions collectively define the fundamental theoretical problem addressed by this treatise. The field of adaptive linear prediction or AR modelling, and the attendant theoretical issues outlined above, have attracted considerable interest from the research community over the past thirty years. Thus we begin our research by reviewing the the works of those before us - for it is in the successes and failures of these efforts that we discover a basis for the solutions that we seek.

1.2 Historical Survey of Relevant Research

The first analytical insight into these crucial performance and design issues was provided by Davisson, [1,7-8], under the assumption that the process to be adaptively predicted is stationary Gaussian with a square integrable power spectral density. Davisson, [1,7], employing the additional (operationally realistic and analytically difficult) constraint that the adaptive predictor is determined from the observation sequence to which it is applied, derived an asymptotically unbiased estimate of the LMS adaptive linear predictor's mean square error as a linear function of the unknown MMSE of the optimal predictor of the same order. Recognizing the significance of this result, Davisson, [8], further derived an asymptotically unbiased estimate of the optimal predictor's MMSE and, combining the two estimators, formulated a method for selecting an adaptive linear predictor's order to minimize its estimated mean square error.

The implications and limitations of Davisson's results and order selection method are discussed further in Chapters 2 and 3.

Akaike, [9-10], apparently unaware of these results, subsequently duplicated and attempted to extend Davisson's Method to stationary AR processes with his publication of the Minimum FPE procedure. In this attempt, Akaike based his arguments in support of the Minimum FPE conjecture on the simplifying assumption that the LMS adaptive predictor is derived from an observation sequence which is independent of the realization to be predicted[6]. Nevertheless, Akaike's theoretical justification of the Minimum FPE conjecture, [10], is critically flawed and the conjecture has been proven invalid (see, for example, [33], [36-37], and Chapters 2, 3).

[6] This assumption is considered unrealistic in typical adaptive prediction applications.

Akaike [11-14], while pursuing the somewhat more general problem of formulating an information theoretic approach to statistical identification, developed an information theoretic criterion (AIC) for statistical model selection on the basis of minimizing the Kullback-Leibler measure of the mean information, provided by a set of observations, $\underline{\mathbf{S}}$, for discriminating between each candidate model $f(\underline{\mathbf{s}}/\underline{\hat{\theta}})$ and the actual probability distribution $f(\underline{\mathbf{s}}/\underline{\theta})$ underlying the data. Akaike's Minimum AIC criterion [12-13] entails selecting the model and its associated Maximum Likelihood (ML) parameter estimates, $\underline{\hat{\theta}}_{ML}$, which minimize

$$AIC = -2\ln\{f(\underline{S}/\underline{\hat{\theta}}_{ML})\} + 2M$$

where M denotes the number of independent parameters defining each model within a set of competing parametric models.

The Minimum AIC criterion is shown, in Chapter 2, to be asymptotically equivalent to Davisson's Method when the process to be predicted is Mth order Gauss-Markov. We examine the limitations of the AIC statistic, and its variants, in Chapter 2.

Another widely cited order selection criterion is found in the work of Parzen [15-16]. Unlike Akaike, Parzen does not assume that the process necessarily admits an autoregressive model (of finite order), but rather seeks to determine the AR model, of finite order, that most closely approximates the (possibly) infinite AR representation of the regular process in question. To this end, Parzen proposes the Criterion Autoregressive Transfer Function (CAT) procedure which suggests that the optimum order, M_0, for the approximating AR model is given by that order M at which

$$CAT(M) = \{\frac{1}{N}\sum_{k=1}^{M}\hat{\sigma}^{-2}[k,\infty]\} - \hat{\sigma}^{-2}[M,\infty]$$

is minimized, where

$$\hat{\sigma}^2[k,\infty] = \frac{F^2(k,N)}{1 - \frac{k}{N}}$$

is the finite sample approximation to the asymptotically unbiased estimate of the kth order optimal predictor's MMSE, as formulated by Davisson (cf. Chapter 2, (2.4)).

Examination of the CAT criterion reveals that, for $N >> 1$ (the usual case),

$$CAT(M) < CAT(M-1) \quad \text{if, and only if,} \quad \hat{\sigma}^2[M,\infty] < \hat{\sigma}^2[M-1,\infty]\,.$$

Thus use of the CAT criterion indeed achieves Parzen's objective of identifying the closest approximating AR model (optimal predictor) in those applications where the LMS residual based statistic, $\hat{\sigma}^2[M,\infty]$, constitutes a reliable estimate of the model's innovations variance. Unfortunately, this statistic, as shown in Chapter 3, can only be relied upon to provide an accurate estimate of $\sigma^2[M,\infty]$ in Gaussian environments.

Parzen does not address the performance of the implied adaptive linear predictor. Moreover, Parzen notes (as have subsequent researchers) that the AIC and CAT criteria typically yield identical AR model order estimates. Hence the CAT criterion does not appear to offer additional potential for extending Davisson's results beyond the Gaussian assumption.

Jones, [17-18], appears to have been the first of many authors to mistakenly attribute Davisson's Method to Akaike. Noting that the Minimum FPE procedure (Davisson's Method) and AIC criterion are asymptotically equivalent and, in tests, "almost invariably select the same order", Jones [17] employs the latter to determine the order of autoregressive spectral estimates for several large sample (i.e., $M/N < 10^{-2}$) biological and meteorological time series. Jones argues, with the results thus obtained, that the procedure provides an objective means of order determination for AR spectrum estimation. Subsequently Jones [18] examines the accuracy of the method's finite sample approximations, $\hat{\sigma}^2[M,\infty]$, and, $\hat{\sigma}^2[M,N]$, (to the asymptotically unbiased estimates of the optimal and adaptive predictor's mean square errors, respectively, cf. section 2.2), via simulation. As a consequence of employing Akaike's restatement of Davisson's Method, Jones is understandably unaware of Davisson's stated conditions (i.e., that the adaptive predictor's order must be much less than the length of its learning period) under which these finite sample based estimates constitute accurate approximations to their asymptotic counterparts. Thus Jones allows the predictor's order, M, to approach the sample size, N, in his simulations and erroneously concludes that the Minimum FPE procedure (Davisson's Method) is biased to yield order estimates which are artificially low. Jones does, however, demonstrate that the method is extendable to the order selection of adaptive linear predictors determined under Burg's Maximum Entropy criterion.

The first variant of the AIC criterion is apparently attributable to Schwartz [19]. Recognizing that the Minimum AIC estimate of model order is inconsistent, Schwartz proposed the BIC statistic

$$BIC = -2\ln\{f\,(\underline{S}/\hat{\underline{\theta}}_{ML})\} + M\ln N$$

and postulated that the Minimum BIC criterion provides a consistent estimator for statistical model selection. The BIC criterion, as a variant of the Minimum AIC procedure, unfortunately inherits the limitations of the latter. These shortcomings are considered in Chapter 2.

In the field of universal coding, Rissanen [20-21] derived the Minimum Description Length (MDL) criterion as a statistical hypothesis test and parameter estimation procedure for determining, from among a set of competing parametric models, that model which permits the shortest possible representation (encoding) of the observation sequence. As such, Rissanen [22] recognized that the MDL criterion constitutes a consistent AR model order selection procedure. Thus, Rissanen claimed that the MDL criterion is superior to the Minimum FPE procedure and the closely related AIC criterion, [13], since the latter two methods yield inconsistent estimates of an adaptive linear predictor's optimal order. Although Rissanen demonstrated that the MDL criterion is, for Gaussian processes, asymptotically equivalent to Davisson's Method, he did not critically examine the applicability or performance of the criterion for non-Gaussian processes. The MDL criterion is shown, in Chapter 2, to be equivalent to the BIC criterion and hence is seen to be subject to the practical limitations of the Minimum AIC procedure.

Hannan and Quinn [23], following the works of Akaike, Schwartz, and Rissanen, propose yet another variant of and alternative to the AIC criterion. Arguing to establish the need for such an alternative, Hannan and Quinn cite the AIC criterion's well known propensity to overestimate the order of an AR process, and the unnecessarily rapid rate of increase, as $N \to \infty$, of the order penalty term $M\ln N$, found in the criterion's strongly consistent BIC and MDL variants. Thus Hannan and Quinn seek a criterion of the AIC form with a cost function order penalty term that increases at the slowest rate possible to effect strong consistency.

To this end, Hannan and Quinn propose the HQ (cost function) statistic

$$HQ = -2\ln\{f\,(\underline{S}\,/\,\hat{\underline{\theta}}_{ML})\} + cM\ln\ln N \quad , \; c > 2$$

and prove that the argument which minimizes it, over $1 \le M \le M_{\max}$, constitutes a strongly consistent estimate of an AR(P) processes' order under the *a priori* assumption that the order in question, $P \le M_{\max}$.

Hannan and Quinn observe that, since the order penalty factor $c\ln\ln N$ of this criterion increases more slowly than the corresponding factor $\ln N$ of the BIC/MDL criteria, the HQ criterion can be expected to underestimate order to a lesser degree than the BIC/MDL criteria in large but finite sample applications. In this respect, Hannan and Quinn suggest that the HQ criterion offers a compromise between the AIC and BIC/MDL criteria.

Hannan extends the HQ criterion to and proves its strong consistency for ARMA(P,Q) processes (again with the *a priori* knowledge that $P \le P_{\max}$, $Q \le Q_{\max}$) in [24]. We therefore consider the HQ criterion further, and examine its limitations as a variant of the AIC procedure, in Chapter 2.

Fuller and Hasza, [25], investigate the MSE performance of LMS adaptive predictors under the assumption that the process to be predicted is Gaussian AR, of known order. Consequently, Fuller and Hasza consider neither the performance degrading effect of order misspecification nor the potential of order selection criteria in adaptive linear prediction. Asymptotically unbiased estimates of the adaptive predictor's MSE are derived when the predictors coefficients are determined from either the realization to be predicted, or a second, independent realization, and are shown to be equal under the assumptions above. Davisson's original contributions, [7], are both acknowledged and independently verified. Fuller and Hasza further demonstrate that LMS estimates of the adaptive predictor's MSE are consistent with and without the stationarity assumption and conclude that stationarity is not a prerequisite in LMS adaptive prediction applications. Unfortunately, Fuller and Hasza do not address the adaptive prediction of non-Gaussian processes.

Bhansali, [26-27], like Akaike, bases his analyses on the assumption that the process to be predicted is stationary, Pth order AR and independent of, but otherwise statistically identical to, the process from which the subject Mth order LMS adaptive predictor's coefficients are estimated. Bhansali admits that "In practice...this independence assumption is unrealistic. The assumption, nevertheless, has been made for the mathematical convenience it affords". Proceeding under this disclaimer and with the additional assumptions that the sample mean estimate, r_M, of the processes' covariance matrix, R_M, remains close to the latter, in a Euclidean (induced matrix) norm sense, for all N sufficiently large (say $> N_o$), i.e.

$$\| r_M R_M^{-1} - I \| \le F_1 < 1 \quad , \; N > N_o \tag{1.9}$$

and that the eighth order moments of the processes' innovations are finite, Bhansali [27] demonstrates that the LMS adaptive predictor converges in mean square to the optimal predictor of the same order at a rate on the order of $N^{-1/2}$. Bhansali, invoking the further constraint that the 16th order moments of the innovations process are finite, derives an estimate of the MSE of the LMS adaptive predictor through terms of order N^{-1}. This estimate is, when $M \ge P$, shown to be consistent with the corresponding result given by Davisson [8].

Bhansali's results, although mathematically elegant and theoretically interesting, have limited practical utility. Indeed, these results are predicated on the condition (1.9) that the sample mean estimates of the processes' second order moments converge deterministically - a behavior which is difficult to ensure; (Bhansali observes that, although the mixing assumption renders this condition plausible, it does not guarantee it). Moreover, Bhansali's estimate of the MSE of the LMS adaptive predictor is given in terms of the unknown second order moments and AR parameters of the process in question. An LMS adaptive predictor order selection criterion, in terms of M, N, and a finite sample approximation to the predictor's estimated MSE, are not formulated.

Bitmead, [28], examines the asymptotic behavior of the recursive LMS adaptive predictor, as popularized by Widrow, [3], under the assumptions that the input process is strictly stationary, ϕ mixing (as defined in [44,46]), and is either uniformly bounded a.e. for all n or has finite moments of all orders. With these constraints, Bitmead demonstrates that the recursive LMS adaptive predictor's coefficients converge in distribution. This conclusion is also shown to result, [29], if a stronger, symmetric mixing condition is substituted for the bounded input assumption. Although Bitmead qualitatively discusses the asymptotic bias and variance of these coefficient estimates, he does not quantify these results nor does he address the mean square prediction error of either the recursive or direct method LMS adaptive predictor. Nonetheless, the relevance of Bitmead's work, to the present research, lies in its incorporation of mixing criteria in place of bounds on the sample paths or moments of the process to be predicted.

Gyorfi, [30], examines the minimal conditions under which almost sure convergence of the LMS adaptive estimates is assured. His research is motivated by the observation that mixing conditions, as invoked by previous authors to ensure strong consistency of the LMS adaptive solution, are "thought to be far from necessary in many cases". To confirm this belief, Gyorfi waives the ergodic assumption on the process in question, requiring only that it is characterized by finite fourth order moments. Under these general conditions, Gyorfi demonstrates that the LMS adaptive estimates converge almost surely to a solution of the generalized Wiener-Hopf equation which is not necessarily either unique or non-random. Regrettably, Gyorfi does not quantify the rate of this convergence, nor does he address convergence in quadratic mean or the asymptotic mean square error performance of the LMS estimate under the stated conditions.

In the context of the present research, consistency (even in the strong sense) is, although theoretically desirable, considered to be of less importance than the mean square error performance manifested by an LMS adaptive mechanism when based on the finite length realization inherent in practical applications. Indeed, in Chapter 4 we demonstrate that the imposition of the consistency constraint, on the adaptive linear predictor's order selection estimate, can in fact degrade the predictor's finite learning period MSE performance from that which otherwise would have been realized. Gyorfi's results are of interest nevertheless, since they guarantee the reassuring property of almost sure convergence without the ergodic assumption.

The most significant contribution towards extending Davisson's results to the adaptive prediction of processes which are not necessarily Gaussian, under the single realization hypothesis, may be found in the work of Kunitomo and Yamamoto, [31][7]. Indeed, this work represents the first attempt to comprehensively analyze the effects of the single, and independent realization hypotheses as well as the abandonment of the Gaussian assumption, on the performance of the LMS adaptive predictor.

In discussing the motivation behind their research, Kunitomo and Yamamoto observe that most (prior) analyses of the LMS adaptive predictor require that its coefficients are determined from a process which is assumed to be an independent realization of the process to be predicted. Kunitomo and Yamamoto define this "conventional assumption" as the *independent case* and designate the more difficult but "realistic" assumption - that the adaptive predictor's coefficients are derived from the process under observation for prediction, as the *dependent case*[8]. With these definitions, Kunitomo and Yamamoto consider the adaptive prediction of a class of stationary, ergodic P th order AR processes and the effects of order misspecification when an M th order LMS adaptive predictor is employed, under the dependent and independent cases; (this class of processes is essentially identical to the class considered by Bhansali, with the exception that the associated innovations process is assumed, in addition to being i.i.d., to be symmetrically distributed about a zero mean). To facilitate their analysis, Kunitomo and Yamamoto assume that

$$E \parallel r_M^{-1} \parallel^{2k} , \quad k = 1, \cdots, k_o \tag{1.10}$$

is bounded for all $N > N_o$ and some k_o. Kunitomo and Yamamoto defend this assumption only by noting that it is similar to the assumption, (1.9), required by Bhansali, and by observing that Fuller and Hasza, [25], have demonstrated its satisfaction by Gaussian processes.

Invoking the additional requirement that the 16th order absolute moments of the processes' innovations are finite, Kunitomo and Yamamoto determine the bias and covariance of the LMS adaptive predictor, through terms of order N^{-1}, and thereby demonstrate that the predictor converges in mean square to the optimal predictor at a rate of $O(N^{-1})$. Kunitomo and Yamamoto, further assuming finiteness of the 32nd order absolute moments of the innovations, derive an expression for the mean square prediction error, under both the dependent and independent cases, through terms of order N^{-1}. Exploiting these results, Kunitomo and Yamamoto cite several specific process examples to prove that the difference between the MSE incurred under the dependent and independent case can be substantial. Thus Kunitomo and Yamamoto conclude that "This raises some question about the conventional assumption of statistical independence between the stochastic process used in estimation and the process [employed] for prediction".

Unfortunately, Kunitomo and Yamamoto's expressions for the adaptive predictor's MSE are, like Bhansali's, based on the restrictive assumption (1.10) and defined in terms of the processes' unknown regression and covariance parameters. Although Kunitomo and Yamamoto suggest that their results might serve as the basis for an order selection criterion for LMS adaptive prediction under the dependent case (citing Akaike's failed attempt to extend Davisson's Method), they do not pursue the matter further.

---

[7] The authors first became aware of this work in March of 1989 as a result of a private communication from Professor C.Z. Wei to Professor L.D. Davisson.

[8] These definitions will, in deference to Kunitomo and Yamamoto, be used throughout the sequel of the present work.

Wei, [32], investigates conditions which ensure strong consistency of a recursive LMS adaptive predictor when the predictor is applied to a (multiple) stochastic regression model whose innovations process is assumed to be a martingale difference sequence (the latter assumption is, clearly, a generalization of the conventional i.i.d. innovations hypothesis). Assuming that the model's true order is known, Wei derives conditions which are sufficient to guarantee almost sure convergence of the predictor's coefficients and its excess sample mean square prediction error. These sufficiency conditions are given in terms of asymptotic properties required of the maximum and minimum eigenvalues of the sample covariance matrix.

Wei subsequently extends these results, [34], to the direct method LMS adaptive predictor. In doing so, Wei observes that, in many applications, the predictor's coefficients are not adaptively updated with each new observation, but rather are updated periodically and applied over several sampling periods. Wei argues that in view of this implementation strategy, the "cumulative cost" as measured by the sample mean square prediction error may be of primary interest. Thus, Wei determines the asymptotic behavior of the excess cumulative cost:

$$\sum_{k=1}^{N} (\Delta \underline{\beta}_{M,N}^{T} \underline{S}_{n-k})^2$$

in probability.

Wei's results are consistent with those of Rissanen (when the process in question is Gaussian) and like Gyorfi's work, provide valuable insight into the asymptotic behavior of the LMS adaptive predictor. However, these results are not directly applicable to the subject of the present study, as discussed above.

The Predictive Least Squares (PLS) criterion, as proposed by Wax, [35], entails selecting the adaptive linear predictor's order to minimize the average "true" prediction error incurred (over the duration of the learning period) by the predictor, within a specified range of admissible orders. Wax defines the "true" prediction error as the sample mean square error induced by an LMS predictor whose coefficients are, at each point within the observation interval, determined by only the "past" data. Thus, the PLS criterion bases the order selection decision on an average of the sample mean square prediction errors resulting from LMS coefficient estimates which are derived, in general, from less than all of the data within the observation interval.

This strategy seems to be inconsistent with one of the fundamental principles of optimal estimation theory - that *all* observation data with the potential to enhance an estimate should in fact be employed to that end. Furthermore, the theoretical justification of the PLS criterion (a consistency proof) is, after Bhansali's work, predicated on the improbable assumption (1.9) that the differences between the sample moments and the corresponding moments of the underlying process are bounded for all sufficiently large observation periods. Although Wax presents a comparison of AR model order estimates obtained, via simulation, from the PLS, AIC, and MDL criteria, he does not address either the comparative performance of Davisson's Method or the MSE performance of the resulting adaptive linear predictor. We therefore remain convinced that a reliable order selection criterion must be based on a performance estimate that accurately reflects the ultimate operation and MSE performance of the LMS adaptive predictor, and consequently, do not consider the PLS statistic further in this capacity.

Finally, Krieger and Masry, [38], derive asymptotic bounds on the "excess" MSE and mean square deviation of an adaptive linear predictor and its coefficients when the predictor's weights are generated by a modified (Widrow gradient search) LMS (MLMS) algorithm, and the input process is either strong mixing or exhibits a maximal correlation coefficient (cf. Chapter 4) that vanishes with increasing intervariate delay. The MLMS algorithm constrains the usual LMS iterative gradient search based coefficient estimates by projecting them onto a predetermined compact subspace of R^M. Thus, the asymptotic bounds on the resulting predictor's excess MSE and the mean square deviation of its coefficients are given in terms of the bound defining the compact solution space and conditioned upon knowledge of the (unknown) minimal eigenvalue of the input processes' covariance function. Unfortunately, Krieger and Masry do not examine the MSE performance of unconstrained adaptive predictors obtained via the direct method (i.e., where the LMS estimate of the predictor's coefficients is obtained by direct solution of (1.4)) nor do they consider an order selection criterion for such a predictor. The contribution of this work is, nevertheless, found in the adoption of the "... fairly weak assumption that the input processes satisfy a strong mixing condition or are asymptotically uncorrelated." Indeed, the strong mixing hypothesis assumes an integral role in the development of the present theory, as will be seen in Chapters 3 and 4.

1.3 Synopsis of the Present Treatise

In view of the above investigations, Davisson's original 1966 observation, [8], that "an unknown factor [in the extension of his results and order selection criterion] is the effect of a departure from the Gaussian assumption..." appears to remain valid. The extension of Davisson's Method, to the adaptive linear prediction of processes which are not necessarily Gaussian, is the principal subject of this treatise.

To this end, we begin with a comprehensive review of the results of Davisson and examine Akaike's attempt to extend these theories, through the Minimum FPE conjecture, to the adaptive prediction of processes which are not necessarily Gaussian. The AIC criterion, and its variants, are also considered as candidates for effecting such an extension. Regrettably, such information theoretic criteria are found to impose restrictions which are unreasonable in adaptive prediction applications, and are thus omitted from further consideration. This review leads to a critical analysis of the assumptions, conclusions, and supporting arguments of the Minimum FPE conjecture. The fallacy in Akaike's arguments is exposed, and several counterexamples to the Minimum FPE conjecture are given.

The preceding analysis not only disproves the Minimum FPE conjecture, but also provides insight into the difficulties in extending Davisson's results, under either the independent or dependent case, without the Gaussian assumption. These impediments are further revealed by theorems which define the mean square error incurred, when applying an LMS adaptive predictor to a process which is not necessarily Gaussian, under both the dependent and independent cases. Indeed, the given theorems clearly indicate that the LMS adaptive predictor's asymptotic MSE, through terms of $O(N^{-1})$, is a function of fourth order moments which are unknown for the process in question. These conclusions are compared to and found to be consistent with the corresponding results of Kunitomo and Yamamoto, [31], for AR and invertible autoregressive-moving average (ARMA) processes.

These findings suggest a possible generalization of Davisson's asymptotically unbiased estimator of the LMS adaptive predictor's MSE and an extension of his order selection criterion which neither requires nor excludes the Gaussian assumption. Thus we consider such an extension to a class of stationary, mixing processes which includes stationary ergodic Gaussians as a subset. We prove that the stated conditions are sufficient to ensure that the generalized estimator is asymptotically unbiased. A finite length realization based approximation of the predictor's MSE immediately follows from the reasoning employed in the derivation of Davisson's original order selection method.

With the support of the above developments, we formally extend Davisson's order selection criterion for LMS adaptive predictors and AR models to the class of stationary processes which meet the prerequisite mixing condition.

The appendices referenced in Chapters 1 through 4, and the proofs of the results given in Chapters 3 and 4 have been omitted due to space limitations. These derivations may be found in [52] and are available from the authors upon request.

The references to the literature, as cited in each chapter, appear following the conclusion of Chapter 4.

Chapter 2
An Assessment of Precursory Works

2.0 Introduction

We begin this chapter, appropriately, with a review of the pioneering works of Davisson, [1, 7-8]. This work appears to have been the first to propose an order selection criterion, for AR modeling and LMS adaptive prediction, on the legitimate theoretical basis of minimizing the estimated mean square prediction error of the model in question. The limitations of Davisson's theories are revisited to provide both the motivation behind and the foundation for the developments of Chapters 3 and 4.

The summary of Davisson's results is followed by an examination of the Minimum Final Prediction Error (FPE) conjecture, as proposed by Akaike [9-10]. With this study, we not only reveal the conspicuous similarity between Akaike's conjecture and Davisson's Method (the former is an exact restatement of, and employs notation which is essentially identical to that found in, the latter), but also begin to consider the difficulties that one encounters when attempting to extend Davisson's results - the Minimum FPE conjecture is shown, as a consequence of the analysis prerequisite to our main result, to be invalid, in Chapter 3.

The examination of the Minimum FPE conjecture leads, quite naturally, to a discussion of four related and widely cited information theoretic based model order selection criteria, beginning with Akaike's generalization of Davisson's Method through the proposal of the celebrated AIC statistic. The Minimum AIC criterion is shown to be asymptotically equivalent to Davisson's Method when the process to be modeled is Gaussian (i.e., in the only environment in which the AIC criterion has been applied). Three variants of the Minimum AIC criterion - the BIC statistic, as developed by Schwartz, [19], Rissanen's Minimum Description Length (MDL) criterion, [20-22], and Hannan and Quinn's HQ criterion [23] are briefly considered. The BIC and MDL criteria are, although derived independently and under differing objectives, shown to be identical. The chapter closes with a short discussion of the rationale behind our decision to extend Davisson's Method for process order identification to the non-Gaussian, dependent case rather than to further examine the potential of the Minimum AIC criterion, or a variant thereof, in this capacity.

2.1 Davisson's Results and Method

Davisson [7-8], provided the first answers to the fundamental questions (c.f. Chapter 1, section 1.1) on the performance and design of the LMS adaptive linear predictor under the premise that the predictor is applied to a zero mean, stationary Gaussian discrete-time stochastic process with unknown, but square integrable, power spectral density (psd), in the dependent case. These answers, and the analytical approach employed to obtain them, are reviewed in the following section as a prologue to our present efforts to extend their applicability.

2.1.1 The Adaptive Linear Predictor's MSE

Davisson's analysis, [7], (1965), of the MSE performance of the LMS adaptive predictor, under the stationary, ergodic Gaussian hypothesis, begins with an examination of the asymptotic behavior of the predictor's coefficients. Thus, Davisson demonstrated that the sample covariances in (1.4) converge in mean square to their respective ensemble averages and that the LMS adaptive predictor weights $\underline{\beta}_{M,N}$ converge in mean square to the optimal LMMSE weights $\underline{\beta}_{M,\infty}$, for all $M < \infty$, as the length of the learning period becomes infinite (i.e., as $N \to \infty$).

With the assurance afforded by these observations, Davisson substituted the Taylor series expansion

$$\Delta \underline{\beta}_{M,N} = R_M^{-1}(r_M R_M^{-1} \underline{G}_M - \underline{g}_M) + \underline{\epsilon}_N \tag{2.1}$$

where

$$E \parallel \underline{\epsilon}_N \parallel^2 = o(\frac{M}{N})$$

of the error $\Delta \underline{\beta}_{M,N}$ in the LMS estimate of the optimal weights in (1.8) to demonstrate that the mean square prediction error of the LMS adaptive predictor is given by

$$\sigma^2[M,N] = \sigma^2[M,\infty](1 + \frac{M}{N}) + o(\frac{M}{N}) \tag{2.2}$$

where

$$\frac{N}{M}\, o(\frac{M}{N}) \to 0 \quad as \quad \frac{N}{M} \to \infty .$$

In discussing the above results, Davisson observed that, since the optimal predictor's error $\sigma^2[M,\infty]$ is a monotonically decreasing function of M while the factor $(1 + M/N)$ is an increasing function of M, the existence of an optimal[1] LMS adaptive predictor order, M_0, is guaranteed, provided $\sigma^2[\infty,\infty] > 0$. Moreover, Davisson concluded that if the behavior of $\sigma^2[M,\infty]$, as a function of M, were known, then (2.2) could be exploited to specify the optimum predictor order M_0 as a function of the learning period N, which is usually determined by the degree of stationarity of the input process and computational complexity constraints.

As seen in the following section, Davisson, motivated by the potential of his discoveries, proceeded to derive an asymptotically unbiased estimate of $\sigma^2[M,\infty]$ and exploited this and the preceding results to propose his LMS adaptive predictor and AR model order selection criterion. (This later work has been overlooked by most researchers - an oversight that is attributable to its publication in a rather esoteric journal [8]).

[1] The terms "optimal" and "optimum" are used in the present context to indicate that order which minimizes the predictor's MSE.

2.1.2 Davisson's Method and its Applicability

Recognizing that the behavior of $\sigma^2[M,\infty]$ is seldom known *a priori* in adaptive applications, and armed with the results of [7], Davisson, [8], (1966) further demonstrated that :

1) The expected value of the sample mean square prediction error is given by

$$E\{F^2(M,N)\} = \sigma^2[M,\infty](1 - \frac{M}{N}) + o(\frac{M}{N}) + \frac{o(M)}{N} \tag{2.3}$$

and thus an asymptotically unbiased estimate of the mean square error, $\sigma^2[M,\infty]$, of the Mth order optimal predictor is realized by

$$\hat{\sigma}^2[M,\infty] = \frac{F^2(M,N)}{1 - \frac{M}{N}} . \tag{2.4}$$

and

2) Combining the results of (2.2) and (2.4) yields an asymptotically unbiased estimate of the LMS adaptive predictor's mean square prediction error, $\sigma^2[M,N]$, given by

$$\hat{\sigma}^2[M,N] = F^2(M,N)\frac{1 + \frac{M}{N}}{1 - \frac{M}{N}} . \tag{2.5}$$

Davisson, on the basis of (2.4) and (2.5), formulated his LMS adaptive linear predictor design method in [8] as follows:

(i) The optimum memory, or order, M_0, is given by the value of M which minimizes $\hat{\sigma}^2[M,N]$ over $1 \le M \le M_{max}$, where M_{max} and N are selected such that $M_{max} << N$.

(ii) The optimum learning period, N_0, is subsequently chosen as the minimum learning period N such that

$$\frac{\hat{\sigma}^2[M_0,N]}{\hat{\sigma}^2[M_{max},\infty]} \le 1 + \epsilon$$

where ϵ represents the acceptable percentage by which the LMS adaptive predictors mean square error is allowed to exceed that of the optimal LMMSE predictor of maximal order. The selection of the optimum learning period N_0, under this criterion, minimizes the degrading effects of potential process nonstationarities as well as the computational complexity of the LMS adaptation algorithm.

(iii) An estimate of the mean square prediction error of the LMS adaptive predictor specified by (i) and (ii) is given by $\hat{\sigma}^2[M_0,N_0]$ as determined by (2.5).

Davisson's Method, as outlined above, is applicable to any stationary Gaussian process of known mean and square integrable psd. The latter requirement constitutes an ergodicity constraint and can be seen to be satisfied by a large class of processes typically found in adaptive applications (e.g., any process whose autocorrelation function is absolutely summable or whose autocorrelation function does not contain undamped periodic components and whose psd is bounded a.e.).

Moreover, it is immediately obvious that Davisson's Method may be applied to adaptive predictors of and AR models for Gaussian processes of unknown mean. The theoretical justification of this claim is found in applying the analytical approach followed in [7-8] to the LMS adaptive predictor obtained from an appropriate modification of (1.4) and (1.6) by the sample means of the process in question.

Although subsequent authors (c.f., Kunitomo and Yamamoto, [31]) have suggested that deviation from the Gaussian assumption has an insignificant effect, in practical applications, on the accuracy of the results above, the most restrictive theoretical constraint in applying applying Davisson's Method remains, nonetheless, to be the assumption that the observation sequence to be adaptively predicted or modeled is Gaussian. The removal of this constraint is the subject of Chapters 3 and 4.

2.2 Akaike's Investigations

Akaike first proposed the Minimum Final Prediction Error (FPE) procedure [9] for selecting the optimum order of an LMS adaptive predictor in 1969 and subsequently attempted to provide a theoretical justification [10] of the procedure in 1970. The Minimum FPE procedure, although derived by employing assumptions and an analytical approach that differed somewhat from Davisson's works, is identical to Davisson's Method.

2.2.1 The Minimum FPE Conjecture

Akaike, viewing the problem as one of fitting an autoregressive (AR) model to an AR stochastic sequence, based his analysis on assumptions which, from application standpoint, are considerably more restrictive than those upon which Davisson derived his method. Akaike predicated his analysis on the unrealistic assumption that the process under observation for the LMS estimation of the adaptive predictor's coefficients is independent of the process to which the adapted predictor is to be applied; i.e., both processes are assumed to be independent realizations of the same P th order stationary AR discrete-time stochastic process

$$S_n = \underline{\beta}_{P,\infty}^T \underline{S}_n + \alpha_{P,\infty} + \epsilon_n \tag{2.6}$$

where the innovations process, $\{\epsilon_n\}$, is assumed to be zero mean i.i.d.. The independent realization assumption allowed Akaike to statistically decouple the adaptive predictor's LMS coefficient estimation errors from the observations to which these errors are applied[2].

With the benefit of these assumptions, Akaike proceeded by defining the Mth order LMS adaptive predictor and its FPE as

$$\hat{S}_n = \underline{\beta}_{M,N}^T \underline{S}_n + \alpha_{M,N}$$

and

$$FPE_M \text{ of } \hat{S}_n = \sigma^2[M,N] \quad \text{as given by (1.7)}$$

[2] The implications of the assumption that the innovations process is not simply white (as it is conventionally defined to be) but independent are discussed in Chapter 3.

where $\underline{\beta}_{M,N}$ is defined by (1.4) with r_M and $\underline{g}_M$ adjusted by the processes' sample means

$$\overline{S}_n = \frac{1}{N}\sum_{k=1}^{N} S_{n-k} \quad \text{and} \quad \overline{\underline{S}}_n = \frac{1}{N}\sum_{k=1}^{N} \underline{S}_{n-k} = (\overline{S}_{n-1},\ldots,\overline{S}_{n-M})^T$$

as

$$r_M = \frac{1}{N}\sum_{k=1}^{N} \underline{S}_{n-k}\,\underline{S}_{n-k}^T - \overline{\underline{S}}_n\,\overline{\underline{S}}_n^T \quad , \quad \underline{g}_M = \frac{1}{N}\sum_{k=1}^{N} S_{n-k}\,\underline{S}_{n-k} - \overline{S}_n\,\overline{\underline{S}}_n$$

and

$$\alpha_{M,N} = \overline{S}_n - \underline{\beta}_{M,N}^T \overline{\underline{S}}_n \ .$$

Employing the above definitions and constraints, with the additional assumption that the order, M, of the LMS adaptive predictor and the actual order, P, of the process on which the predictor operates are equal, Akaike showed that, asymptotically, as the learning period becomes infinite:

$$FPE_M = \sigma^2[M,\infty](1 + \frac{M+1}{N}) \tag{2.7}$$

and that an asymptotically unbiased estimate of FPE_M is given by

$$\hat{FPE}_M = F^2(M,N)\frac{1 + \frac{M+1}{N}}{1 - \frac{M+1}{N}} \ . \tag{2.8}$$

In view of these results, Akaike, employing reasoning identical to Davisson's rationale, proposed the Minimum FPE procedure for selecting the optimum LMS adaptive predictor/AR model order, M_0, as selecting M_0 such that

$$M_0 = \arg \min_{1 \le M \le M_{max}} \hat{FPE}_M \ . \tag{2.9}$$

Arguing asymptotically in an attempt to complete the justification of the Minimum FPE procedure, Akaike demonstrated that if the order, P, of the AR process (2.6) under observation, and the order, M, of the LMS adaptive predictor under consideration are not equal, then:

i) If $M < P$, then $\lim_{N\to\infty} P\{\hat{FPE}_M - \hat{FPE}_P > 0\} = 1$

and

ii) If $M > P$, then $\lim_{N\to\infty} P\{\hat{FPE}_M - \hat{FPE}_P > 0\} = 1 - P\{\chi_{M-P}^2 > 2(M-P)\}$,

where χ_{M-P}^2 is Chi-square distributed with $M-P$ degrees of freedom. Thus, although the Minimum FPE procedure estimate, M_0, of the LMS adaptive predictors optimum order is not a consistent estimate of P, Akaike maintained that the asymptotic probability of adopting an order M_0, which exceeds the optimum order P:

$$P\{\chi_{M_0-P}^2 > 2(M_0-P)\} < 0.2$$

is "not necessarily intolerable for practical applications."

Comparing (2.7), (2.8) and (2.9) to (2.2) and (2.5) and Davisson's definition of M_0, respectively, and recognizing that the addition of 1 to the LMS adaptive predictor's order M in Akaike's equations reflects his allowance for unknown mean processes, we see that Akaike's Minimum FPE procedure is identical to Davisson's Method. Unfortunately, the validity of this apparent extension of Davisson's order selection criterion is disproved in Chapter 3, under the conditions for which it is proposed.

2.2.2 The AIC Criterion and its Variants

Akaike developed his now famous information theoretic criterion (AIC), [11-13], for statistical model selection while pursuing the generic problem of formulating an information theoretic approach to statistical K-ary hypothesis testing. The AIC statistic supports statistical model identification on the basis of asymptotically minimizing the Kullback-Leibler measure of the mean information, provided by a set of observations, $\underline{S}$, for discrimination between each candidate model $f\,(\underline{s}/\hat{\underline{\theta}}\,)$ and the actual probability distribution $f\,(\underline{s}/\underline{\theta})$ underlying the data. Thus, Akaike's Minimum AIC criterion [12-13] entails selecting the model and is associated Maximum Likelihood (ML) parameter estimates, $\hat{\underline{\theta}}_{ML}$, which minimize the statistic:

$$AIC = -2\ln\{f\,(\underline{S}\,/\hat{\underline{\theta}}_{ML}\,)\} + 2M$$

where M denotes the number of independent parameters defining each model over a set of competing (and perhaps dissimilar) parametric models.

In applying the Minimum AIC criterion, Akaike [13] claimed that if the candidate models were restricted to those arising from stationary Pth order AR processes (2.6), under the "slightly weaker assumption of normality" (i.e., stationary Pth order Gauss-Markov processes), that:

(i) The ML estimates of the AR coefficients and innovations variance (i.e., the model parameters) are (with $M = P$) given by the LMS estimates (1.4) and the corresponding sample mean square prediction error (1.5), respectively, and

(ii) Asymptotically, the Minimum AIC criterion is equivalent to the Minimum FPE Procedure (Davisson's Method, under the stated constraints); i.e., as $N \to \infty$:

$$M_0 = \arg \min_{1 \le M \le M_{\max}} AIC(M) = \arg \min_{1 \le M \le M_{\max}} \hat{\sigma}^2[M,N]$$

where $\hat{\sigma}^2[M,N]$ is given by (2.5).

The first asymptotic relationship is well known while a proof of the second is given in Appendix 2A.

Akaike did not, however, address the application of the AIC criterion to the order selection of non-Gaussian AR processes.

Schwartz, [19], proposed the BIC statistic:

$$BIC = -2\ln\{f\,(\underline{S}\,/\hat{\underline{\theta}}_{ML}\,)\} + M\ln N$$

as an alternative to the AIC cost function for statistical model selection. As such, the BIC criterion represents an asymptotic approximation (truncated expansion) of the Bayes estimation rule that results from the adoption of a uniform cost function and the assumption of a conditional probability density function (pdf), on the observations, from the Koopman-Darmois family and a class of a priori distributions, on each models parameters, which are bounded above and below (away from zero). The uniform cost function affixes a constant penalty for any and all model errors and, of course, leads to a Bayes rule which is a Maximum A posteriori Probability (MAP) criterion.

In claiming the superiority of the BIC over the AIC, Schwartz notes that, when eight or more observations ($N \geq 8$) are employed, the BIC clearly demonstrates the tendency to favor model orders which are lower than those preferred by the AIC. This observation lends intuitive credibility to the fact that the BIC statistic, unlike the AIC criterion which is known to be inconsistent (as shown in the preceding section, the Minimum AIC criterion exhibits the propensity to over-parameterize the model), has been shown, [39], to be a consistent estimator of model order. Although Schwartz questions the optimality of the AIC criterion, he examines neither the application nor the performance of either the AIC or BIC criterion.[3]

Rissanen [20-22] approached the problem of statistical model identification from the novel perspective of universal coding, with the attendant conviction that the selected model must permit the shortest possible (coded) representation of the observed sequence. In doing so, Rissanen derived the Minimum Description Length (MDL) statistic:

$$MDL = \min_{M}\{ -\log\{f\,(\underline{S}\,/\hat{\underline{\theta}}_{ML})\} + \frac{M}{2}\log N \}$$

where

$$M = \text{parametric dimension of the model } f\,(\underline{S}\,/\underline{\theta})$$

as a measure of the information contained in, and a lower bound to the mean code length realized by any universal coder with respect to the class of models $f\,(\underline{S}\,/\underline{\theta})$ under consideration for, the observed sequence $\underline{S}$. Hence, Rissanen claimed, the MDL model estimate allows the most efficient coding of the observation sequence over all universal codes.

In [22], Rissanen applies the MDL criterion to derive a lower bound on the mean square error of adaptive linear predictors of Gaussian ARMA processes. This lower bound is shown to be achievable by LMS adaptive predictors in the special case of Gaussian AR processes with the application of Davisson's result (2.2)[4]. Rissanen does not characterize the adaptive mean square prediction error of non-Gaussian processes with the MDL criterion and in fact questions the relevance of MSE as a measure of a model's effectiveness in such situations.

Although Rissanen independently derived the MDL criterion, comparison of the BIC and MDL criteria proves that the model estimates which they produce are identical. This, when coupled with the seemingly disparate objectives under which these criteria were derived, leads one, initially, to ponder their potential significance. However, closer inspection reveals that Rissanen employs the Kullback-Liebler measure of the information provided by the observation sequence to quantify the inherent redundancy of the implied code. Thus minimizing encoded description length is equivalent to minimizing the Kullback-Liebler information distance as proposed by Akaike and Schwartz.

---

[3] Akaike's responds to this critique in [14], where he claims optimality of the AIC criterion, in a Bayesian-Minimax sense, under the assumption that the data is conditionally distributed as independent Gaussian variates.

[4] Rissanen acknowledges that "...[Davisson's Method] amounts to the first statement of the FPE criterion and the closely related AIC criterion for order estimation."

Recognizing this common underlying objective in the derivations of the AIC criterion and its variants, it is not surprising that such procedures can be expressed as special cases of a generic Information Theoretic Criterion (ITC). This ITC specifies that the optimal model order, M_0, is given by

$$M_0 = \arg \min_{1 \le M \le M_{\max}} ITC(M)$$

where

$$ITC(M) = -2\ln\{f(\underline{S}/\hat{\underline{\theta}}_{ML})\} + p(N)M \tag{2.10}$$

and $p(N)$ is a nondecreasing (order penalty factor) function of N. Obviously, $p(N) = 2$ and $\log N$ for the AIC and BIC/MDL cost functions, respectively.

If it is further assumed that the process to be modelled is a Gaussian AR(M) process (i.e., the only and often implicit condition under which the AIC criterion and its variants have been invoked) then the ITC cost function becomes

$$ITC(M) = N\ln F^2(M,N) + p(N)M \tag{2.11}$$

where $F^2(M,N)$ denotes the LMS residual or sample mean square prediction error as defined in (1.5).

Hannan and Quinn [23] question the performance of the AIC and BIC/MDL variants of (2.10) or (2.11). Noting the AIC criterion's well known lack of consistency (even in the weak sense), they observe further that the BIC/MDL variant's order penalty factor $p(N) = \ln N$ increases at a rate greater than that which is necessary to guarantee the strong consistency of such variants. Thus Hannan and Quinn intuitively argue that use of the BIC/MDL criteria could, in large sample applications, induce underestimation of model order.

Motivated by such concerns, Hannan and Quinn propose the order penalty factor $p(N) = c\ln\ln N$, $c > 2$. The resulting Hannan-Quinn order selection cost function follows, from (2.10), as

$$HQ = -2\ln\{f(\underline{S}/\hat{\underline{\theta}}_{ML})\} + Mc\ln\ln N \quad , \quad c > 2 \tag{2.12}$$

which, with the implicit assumption of a Gaussian likelihood function, becomes

$$HQ = N\ln F^2(M,N) + Mc\ln\ln N \quad , \quad c > 2 \,. \tag{2.13}$$

Proceeding under the assumption that the process to be modelled is AR(P) (as defined in (2.6)), with an innovations process $\{\epsilon_n\}$ which is a martingale difference sequence constrained such that $E(\epsilon_n^2/M_{-\infty}^{n-1}) = \sigma^2[P,\infty]$ and $E\epsilon_n^4 < \infty$, where $M_{-\infty}^n$ is the σ algebra generated by $\{\cdots S_{n-1}, S_n\}$, Hannan and Quinn prove that the order estimate defined by

$$M_0 = \arg \min_{1 \le M \le M_{\max}} HQ \,, \tag{2.14}$$

with HQ given by (2.13), constitutes a strongly consistent estimate of the processes' order, P, given *a priori* knowledge that $P \le M_{\max}$. Moreover, Hannan and Quinn show that, under the given conditions, $p(N) = c\ln\ln N$ with $c > 2$ represents the order penalty factor that increases, with N, at the slowest rate possible to effect almost sure convergence of an ITC order estimate. Hannan [24] further demonstrates that the HQ criterion (2.14) yields a strongly consistent estimate of the orders P and Q of an ARMA(P,Q) process under a hypothesis analogous to that adopted in [23], and summarized above. Nevertheless, the HQ criterion clearly represents another, albeit apparently improved, variant of the AIC procedure or special case of the ITC criterion.

Neglecting the deceptively attractive theoretical arguments offered in their derivation or defense, reflection on the AIC, BIC/MDL, and HQ model selection criteria reveals that such procedures constitute intuitively simple modifications of the Generalized Likelihood test found in classical K-ary hypothesis testing theory. These modifications, as seen in (2.10), entail the subtraction of a penalty function which is proportional to the model's parametric dimension ($2M$, $M \ln N$, and $Mc \ln\ln N$ for a model defined by M parameters in the AIC, BIC/MDL, and HQ procedures, respectively). Thus, selecting the model with the maximum *modified* Generalized Likelihood function ensures, as with the application of the AIC, BIC/MDL and HQ criteria, adherence to the "principle of parsimony" in statistical model construction.

Unfortunately, use of the AIC criterion and its variants, as in the application of the Generalized Likelihood test of K-ary decision theory, requires that the modeler postulate K candidate conditional pdfs to potentially explain the observations in question[5]. In the context of classical decision theory, it is assumed (and probable) that the modeler has some basis for such prognostication, including, but not limited to, insight into the physical phenomenon or mechanisms which are anticipated to give rise to the data. However, in adaptive processing applications this insight is seldom if ever available. Indeed, it is precisely the *lack* of such insight, in these applications, that forces the designer to resort to an adaptive approach.

Thus, in spite of their theoretical appeal and apparent popularity, we are faced with a paradox in the use of ITC criteria in adaptive applications. If we knew, or had reason to believe that we could reliably postulate, the potential conditional probability laws governing the observations in question, why then would we not simply employ an optimal approach (e.g., based on the Generalized Likelihood test[6]) to define the necessary processing. On the other hand, if we have no basis for postulating the functional form and parametric dependence of the required conditional probability laws, as is often the case in adaptive applications, use of the AIC criterion or one of its variants would force us to make *a priori* inferences[7] about such laws.

Furthermore, questions on both the model order estimation accuracy and the implied adaptive linear predictor's MSE performance remain unanswered for the AIC, BIC/MDL, and HQ criteria in environments where the Gaussian assumption is clearly inappropriate. Indeed, Hannan [39], after demonstrating consistency of the BIC/MDL criterion order estimates notes that "...in practice [such criteria] must be used via *ad hoc* choices (such as Gaussian Likelihoods).". Shibata (in a commentary following Hannan) reinforces this point by observing that "...consistency of order estimation [and] the mean squared error of the resulting parameter estimates are not compatible." (c.f. statement on consistency in Chapter 1, section 1.2). These questions become doubts when, in Chapter 3, we prove that the asymptotic equivalent of the AIC criterion in the Gaussian case (the FPE conjecture) is in fact invalid in non-Gaussian applications.

---

[5] Akaike [13] implicitly acknowledges this inherent limitation with the claim that the AIC criterion "...is at present the only procedure applicable to every situation where the likelihood can be properly defined...".

[6] Albeit, as modified with a penalty proportional to model dimensionality.

[7] This is, under the present circumstances, a euphemism for speculation.

Therefore, in view of the unsubstantiated performance of the AIC criterion and its variants in the non-Gaussian case, as well as the limited practical utility of such criteria in adaptive applications, we elected to pursue an extension of Davisson's Method that promised to be free of such deficiencies. This extension, as proposed and demonstrated in Chapter 4, necessitates neither the Gaussian assumption nor speculation on the conditional pdf that governs the observed data.

In closing this chapter we note that the AIC criterion and its variants have commanded considerable attention from the research community in recent years due, primarily, to their intellectually appealing theoretical foundations. In view of this, and the relevance of such criteria to our present work, we felt that they deserved consideration in the present treatise. However, we neither intended nor claim to provide a thorough treatment of information theoretic based order selection criteria here.[8]

[8] The literature is replete with papers on the subject; c.f. Hannan, [39], for an comprehensive survey.

Chapter 3
Impediments in Extending Davisson's Method

3.0 Introduction

As revealed in Chapter 2, the Minimum FPE conjecture bears a remarkable resemblance to and appears, upon first impression, to extend Davisson's Method to the class of stationary AR processes. Intrigued by this discovery[1] and its attendant possibilities, we were compelled to critically examine the conjecture's premises, derivation, and supporting theoretical arguments. This examination proved enlightening, for the premises and theoretical arguments behind the FPE conjecture were found to be unsustainable and the conjecture was subsequently proven invalid.

Our rationale for presenting the detailed results of this examination here is twofold. First and foremost, without conclusive proof of the Minimum FPE conjecture's fallacy, our research could be viewed as unnecessary and perhaps even redundant. Secondly, with the detailed scrutinization of the conjecture's failure, we expose the principal obstacles to the extension of Davisson's Method beyond the limitations imposed by the adoption of the Gaussian assumption in its original derivation.

To this end, we begin with an overview of the fallacious reasoning behind the Minimum FPE conjecture. Having demonstrated the flaws in its theoretical support, we proceed to disprove the conjecture in both the *independent case* (under which its validity is claimed) and the *dependent case* hypotheses of adaptive linear prediction and AR modelling[2].

The Minimum FPE conjecture is first disproved in the *independent* case, under which it is derived, with two counterexamples. Subsequently, it is shown that the conjecture is false, even with the Gaussian assumption, in this case. We provide a general spectral formula for the MSE of the adaptive linear predictor in the *independent* case, under the premise of a relatively benign mixing condition, as further evidence of the conjecture's failure. These results are compared with and verified by the independent developments of Kunitomo and Yamamoto [31].

We briefly discuss the limited practical utility and unrealistic aspects of the *independent* case assumption in typical adaptive linear prediction applications to justify the necessity of an analysis of the Minimum FPE conjecture in the *dependent* case. The counterexamples given under the *independent* case analyses are revisited to demonstrate that the conjecture can produce unbounded errors, in typical adaptive linear prediction environments, where *dependent* case hypothesis conditions prevail. This conclusion is further strengthened by a theorem providing a general

[1] The authors were first made aware of Akaike's publications [9-10] by Dr. Jorma Rissanen in August 1983 (in a private letter to Professor Davisson).

[2] The terms *dependent case* and *independent case*, as originated by Kunitomo and Yamamoto [31], are defined below.

spectral formula for the adaptive linear predictor's MSE in the *dependent* case. Somewhat surprisingly, the functional form of the adaptive linear predictor's MSE is shown, in the *dependent* case, to be a special case of the form obtained under the *independent* case. As in the *independent* case, our results are found to be in agreement with the work of Kunitomo and Yamamoto.

With the insight gained from the preceding study we identify a potential approach to the legitimate extension of Davisson's Method in the adaptive linear prediction of processes, which are not necessarily Gaussian, in the *dependent* case. This approach is formulated and analyzed in Chapter 4.

Before proceeding with the main developments of this chapter, we digress to reiterate (from Chapter 1) the definitions of the *dependent* and *independent* case hypotheses in adaptive linear prediction.

Definitions:

(i) The hypothesis allowing that the adaptive linear predictor's regression coefficients are estimated on the basis of the (single) observation sequence to which they are applied is said to be the *dependent case*.

(ii) The premise requiring that these regression coefficients are determined by a (second) realization that is independent of, but otherwise statistically identical to, the observation sequence to which they are applied is said to be the *independent case.*

3.1 The Fallacy of the Minimum FPE Conjecture

The Minimum FPE conjecture is shown to be invalid, both with and without the Gaussian assumption (on the process to be predicted), in the independent case under which it is derived. Moreover, the conjecture is not only proven false, but also shown, potentially, to be severely misleading typical adaptive linear prediction environments where the dependent case hypothesis is necessitated. These proofs are based on specific counterexamples as well as general theorems. We begin our development with an examination of the arguments that purportedly support the Minimum FPE conjecture.

3.1.1 Akaike's Erroneous Supporting Arguments

Akaike's attempted theoretical justification [10][3] of the Minimum FPE conjecture is examined and shown to be both strained and unfounded. This realization provoked our original suspicion of and subsequent investigation into the conjecture's validity.

We begin by applying Akaike's independent case assumption (i.e., that the LMS adaptive predictor, and the observations on which it operates, are derived from statistically independent realizations of the Pth order AR process (2.6) to be predicted) to the second and third terms on the right-hand side of (1.8), to obtain

$$E(S_n - \underline{\beta}_{M,\infty}^T \underline{S}_n)\Delta\underline{\beta}_{M,N}^T \underline{S}_n = E(S_n - \underline{\beta}_{M,\infty}^T \underline{S}_n)\underline{S}_n^T E\,\Delta\underline{\beta}_{M,N}$$

and

$$E(\Delta\underline{\beta}_{M,N}^T \underline{S}_n)^2 = E\,\Delta\underline{\beta}_{M,N}^T R_M \Delta\underline{\beta}_{M,N} = Trace\ R_M E\,\Delta\underline{\beta}_{M,N}\Delta\underline{\beta}_{M,N}^T \ ,$$

respectively.

[3] Appeared one year subsequent to the initial publication of the conjecture [9].

Furthermore, the Orthogonality Principle implies that

$$E(S_n - \underline{\beta}^T_{M,\infty}\underline{S}_n)\underline{S}_n \equiv \underline{0}$$

thus, under the independent case hypothesis, (1.8) becomes

$$\sigma^2[M,N] = \sigma^2[M,\infty] + Trace\ R_M E\, \Delta\underline{\beta}_{M,N} \Delta\underline{\beta}^T_{M,N} \ . \tag{3.1}$$

Invoking Diananda's Central Limit Theorem (for the output of a linear system driven by an m-dependent stochastic input) [41] and employing the assumption of an i.i.d. innovations sequence, Akaike observes that the the random vector $\sqrt{N}\ \Delta\underline{\beta}_{M,N}$ converges in distribution to a zero mean, M dimensional jointly Gaussian distribution, defined with a covariance matrix $\sigma^2[M,\infty]R_M^{-1}$, as $N \to \infty$. It is at this juncture that Akaike makes the first critical error. Given that $\sqrt{N}\ \Delta\underline{\beta}_{M,N}$ converges in distribution to $N(\underline{0}\ ,\ \sigma^2[M,\infty]R_M^{-1})$, Akaike surmises that

$$E\, \Delta\underline{\beta}_{M,N} \Delta\underline{\beta}^T_{M,N} = \frac{1}{N}\sigma^2[M,\infty]R_M^{-1}$$

(not true in general) and with (3.1) mistakenly obtains (2.7). Similarly flawed reasoning is used to fabricate an asymptotically unbiased estimate of the mean square prediction error, of the Mth order optimal predictor, in an attempt to complete the justification of the Minimum FPE conjecture.

Clearly, the fact that $\sqrt{N}\ \Delta\underline{\beta}_{M,N}$ converges in distribution to $N(\underline{0}\ ,\ \sigma^2[M,\infty]R_M^{-1})$ does not necessarily imply that the covariance of $\sqrt{N}\ \Delta\underline{\beta}_{M,N}$ tends to $\sigma^2[M,\infty]R_M^{-1}$ (cf. Cramer [42], pp. 213-214 and 366-367). Therefore, Akaike's arguments fail to prove that the Minimum FPE conjecture constitutes a reliable extension of Davisson's Method to stationary AR processes. Nonetheless, one could speculate, at this point, that the conjecture might in fact be true, despite the faults in its proof. The following developments prove conclusively that this is not the case.

3.1.2 Counterexamples to the Conjecture

We establish a theoretical foundation that is prerequisite to the remainder of this treatise before proceeding with the development of the subject counterexamples. To this end, we must consider the following:

Definitions:

(i) Let $\{S_n\}_{n=-\infty}^{+\infty}$ be a stationary discrete parameter stochastic process defined on the probability space (Ω,B,P), and denote: M_n^{n+k}, for $k \geq 0$, $M_{-\infty}^n$, and $M_n^{+\infty}$ as the σ algebras generated by the random variables $\{S_n, \ldots, S_{n+k}\}$, $\{\ldots,S_{n-1},S_n\}$, and $\{S_n, S_{n+1},\ldots\}$, respectively.

(ii) The stationary process $\{S_n\}_{n=-\infty}^{+\infty}$ is said to be *strong mixing* or to satisfy a *strong mixing* condition[4] if for all $k \geq 0$,

$$\sup_{A \in M_{-\infty}^n,\ B \in M_{n+k}^{+\infty}} | P(A \cap B) - P(A)P(B) | = \alpha_k$$

where $\alpha_k \downarrow 0$ as $k \to \infty$.

[4] This condition is attributed to Rosenblatt [43], who, in 1955, proposed it as a prerequisite to his central limit theorem for dependent random sequences.

Clearly then, given a strong mixing process, we have

$$| P(A \cap B) - P(A)P(B) | \leq \alpha_k$$

for all $A \in M_{-\infty}^{n}$, $B \in M_{n+k}^{+\infty}$, where α_k is referred to as the *strong mixing* (or *dependence*) *coefficient* of the process.

Although this mixing condition may initially appear to be quite formidable, its satisfaction[5] is an immediate consequence of the asymptotic independence of the processes' future from its past and present. Moreover, the strong mixing coefficient α_k can be viewed as a measure of the dependence between the σ algebras $M_{-\infty}^{n}$ and $M_{n+k}^{+\infty}$ and hence between any measurable functions defined on them. Indeed, the rate at which α_k monotonically decreases quantifies the rate at which samples from the process become asymptotically independent. This latter fact becomes more evident with the following lemma, which we state without proof (cf. Hall and Heyde [44] pp. 276-278 for the required proof).

Lemma 3.1 (Davydov)

Given random variables X and Y which are $M_{-\infty}^{n}$ and $M_{n+k}^{+\infty}$ measurable respectively, with $E | X |^{2+\delta} < \infty$, $E | Y |^{2+\delta} < \infty$, then

$$| EXY - EX\, EY | \leq 8\alpha_k^{\frac{\delta}{2+\delta}} \{ E | X |^{2+\delta} E | Y |^{2+\delta} \}^{\frac{1}{2+\delta}} . \tag{3.2}$$

for any $\delta > 0$.

We now employ Lemma 3.1 to derive the following results, which prove to be quite useful in subsequent developments.

Lemma 3.2

Suppose $\{S_n\}_{n=-\infty}^{+\infty}$ is a zero mean[6] stationary process satisfying the strong mixing condition with mixing coefficient α_k. If

$$\sum_{k=1}^{\infty} \alpha_k^{\frac{\delta}{2+\delta}} < \infty \quad \text{and} \quad E | S_0 |^{2+\delta} < \infty \quad \text{for some} \quad \delta > 0 , \tag{3.3}$$

then

(i) The process satisfies an ergodic mean theorem with

$$E \{ \frac{1}{N} \sum_{k=1}^{N} S_k \}^2 = O(\frac{1}{N}) . \tag{3.4}$$

(That is, the sample mean converges in mean square to the ensemble mean) and

(ii)

$$\sum_{m=-\infty}^{+\infty} | R_m |^t < \infty \quad \text{for all} \quad 1 \leq t < \infty \tag{3.5}$$

where $R_m = ES_0 S_m = R_M(i,j)$ for all $1 \leq i, j \leq M$ such that $| i - j | = m$. (In the sequel, the subscript M (which indicates model order) is omitted for brevity whenever ambiguity will not result.)

[5] We examine the strong mixing condition and characterize the class of processes meeting it in Chapter 4.

[6] The zero mean assumption is not necessary for the theory, but is adopted to be consistent with the remainder of this treatise.

Clearly, since absolutely convergent series are also summable, (ii) of Lemma 3.2 implies

$$\sum_{m=-\infty}^{+\infty} R_m^t < \infty \tag{3.9}$$

for any finite $t \geq 1$. Although (3.9) is sufficient for our needs, we chose, nevertheless, to demonstrate the stronger result (ii) in the lemma.

The strong mixing assumption is invoked, and Lemmas 3.1 and 3.2 are employed, in the developments which follow to facilitate the evaluation of the mean and covariance of several nonlinear multivariate functions of process sample moments through terms of $o(\frac{M}{N})$.

The first such development entails an analysis of a truncated Taylor series expansion of the error, $\Delta\underline{\beta}_{M,N}$, in the Mth order adaptive linear predictor's LMS estimate of the coefficients of the optimal predictor of the same order. Our primary interest centers on the asymptotic behavior of the second order moments of the Taylor series remainder in this expansion. This behavior is determined by the following lemma.

Lemma 3.3

Given that the process to be modelled is stationary and strong mixing, then the error in the Mth order adaptive linear predictor's LMS estimates of the optimal regression coefficients (of the same order), $\Delta\underline{\beta}_{M,N} = \underline{\beta}_{M,\infty} - \underline{\beta}_{M,N}$, may be expressed as:

$$\Delta\underline{\beta}_{M,N} = R_M^{-1}(r_M R_M^{-1} \underline{G}_M - \underline{g}_M) + \underline{\epsilon}_N \quad a.e. \tag{3.10}$$

where

$$E \, \| \underline{\epsilon}_N \|^2 = o(\frac{M}{N})$$

and R_M, $\underline{G}_M$ and r_M, $\underline{g}_M$ are defined in (1.2) and (1.4), respectively.

The derivation of (3.10) is based on a generalization of Cramér's result (cf. [42], Theorem 27.7.3, pp. 353-356) on the asymptotic mean square behavior of the remainder resulting from a truncated Taylor series expansion of a nonlinear function of process sample moments. This relationship was first employed in the present context by Davisson [7] where the stationary ergodic Gaussian assumption was exploited to prove its validity. The proof of Lemma 3.3 under the current hypothesis is rather lengthy and thus is given in Appendix 3A.

The mean square prediction error of the adaptive linear predictor, can be determined, in the dependent and independent cases, by exploiting the inherent implications of Lemmas 3.1, 3.2, and 3.3. We begin with the independent case.

Theorem 3.1

Let $\{S_n\}_{n=-\infty}^{+\infty}$ be a zero mean, stationary process satisfying the strong mixing condition with mixing coefficient α_k. Further, assume that

$$\sum_{k=1}^{\infty} \alpha_k^{\frac{\delta}{2+\delta}} < \infty \quad \text{and} \quad E\,|S_n|^{2(2+\delta)} < \infty \quad \text{for some } \delta > 0 \,. \tag{3.11}$$

Then the mean square prediction error, $\sigma^2[M,N]$, of the Mth order adaptive linear predictor is, in the *independent* case, given by:

$$\sigma^2[M,N] = \sigma^2[M,\infty] + \frac{1}{N^2}\sum_{k=1}^{N}\sum_{l=1}^{N} EZ_{n-k}\, Z_{n-l}\, \underline{S}_{n-k}^T R_M^{-1} \underline{S}_{n-l} + o\left(\frac{M}{N}\right) \tag{3.12}$$

where $Z_n = S_n - \underline{\beta}_{M,\infty}^T \underline{S}_n$ is the error incurred by the Mth order optimal predictor.

Reflecting on (3.12) reveals that if the Mth order optimal predictor's MSE, $\sigma^2[M,\infty]$, and the fourth order moments, $\{EZ_0 Z_m\, \underline{S}_0^T R_M^{-1} \underline{S}_m,\ m=-N+1, \ldots, N-1\}$, were indeed known, then an asymptotically unbiased estimate of the Mth order adaptive linear predictor's MSE would be given by

$$\hat{\sigma}^2[M,N] = \sigma^2[M,\infty] + \frac{1}{N^2}\sum_{m=-N}^{N} (N - |m|) EZ_0 Z_m\, \underline{S}_0^T R_M^{-1} \underline{S}_m \tag{3.16}$$

for stationary strong mixing processes in the independent case. This result can, for the purposes of comparison, be expressed as

$$\hat{\sigma}^2[M,N] = \sigma^2[M,\infty] \left[1 + \frac{\mu(M,N)}{N} \right] \tag{3.17}$$

where $\mu(M,N)$ is defined as the arithmetic mean in (3.16) as normalized by $\sigma^2[M,\infty]$.

Fortunately, as seen in the developments which follow, an asymptotically unbiased estimator of $\sigma^2[M,\infty]$ does exist, albeit under the presumption of knowledge of the fourth order moments $EZ_0 Z_m\, \underline{S}_0^T R_M^{-1} \underline{S}_m$.

The following theorem determines the expected value of the LMS adaptive predictor's sample mean square prediction error through terms of order $O\left(\frac{M}{N}\right)$. Unlike Theorem 3.1 (which pertains only to the independent case), the result given here is applicable in both the dependent and independent cases.

Theorem 3.2

Let $\{S_n\}_{n=-\infty}^{+\infty}$ be a stationary process satisfying the conditions of Theorem 3.1. Then

$$E\ F^2(M,N) = \sigma^2[M,\infty] - \frac{1}{N^2}\sum_{k=1}^{N}\sum_{l=1}^{N} EZ_{n-k}\, Z_{n-l}\, \underline{S}_{n-k}^T R_M^{-1} \underline{S}_{n-l} + o\left(\frac{M}{N}\right) \tag{3.18}$$

where $F^2(M,N)$ is the sample mean square prediction error, (LMS fitting error or residual) of the adaptive linear predictor, as defined by (1.5).

Examining (3.18), we see that an asymptotically unbiased estimate of the Mth order optimal predictor's MSE is realized by

$$\hat{\sigma}^2[M,\infty] = F^2(M,N) + \frac{1}{N^2}\sum_{m=-N}^{N} (N - |m|) EZ_0 Z_m\, \underline{S}_0^T R_M^{-1} \underline{S}_m \tag{3.23}$$

or, in the alternative, given by

$$\hat{\sigma}^2[M,\infty] = F^2(M,N) \left[1 - \frac{\mu(M,N)}{N} \right]^{-1} \tag{3.24}$$

with $\mu(M,N)$ as defined in (3.17), for stationary strong mixing processes in both the dependent as well as the independent cases. Once again, however, knowledge of the fourth order moments $EZ_0 Z_m\, \underline{S}_0^T R_M^{-1} \underline{S}_m$ is required to completely determine these estimators.

With the benefit of Theorems 3.1 and 3.2, and the observations following them, we are in a position to derive estimates for $\sigma^2[M,N]$. Thus, combining (3.16) and (3.23) we obtain an asymptotically unbiased estimate of the mean square error

$$\hat{\sigma}^2[M,N] = F^2(M,N) + \frac{2}{N^2} \sum_{m=-N}^{N} (N - |m|) E Z_0 Z_m \underline{S}_0^T R_M^{-1} \underline{S}_m \tag{3.25}$$

incurred by the Mth order adaptive linear predictor in the independent case.

An alternate but nonetheless equivalent estimate of $\sigma^2[M,N]$ is obtained from (3.17) and (3.24) as

$$\hat{\sigma}^2[M,N] = F^2(M,N) \left[\frac{1 + \dfrac{\mu(M,N)}{N}}{1 - \dfrac{\mu(M,N)}{N}} \right] \tag{3.26}$$

and constitutes the basis for a comparative analysis of the FPE statistic's accuracy and hence a test of the validity of the Minimum FPE conjecture as well.

In view of (3.12), (3.16-3.17) and (3.25-3.26), it is clear that the MSE of the adaptive linear predictor is dependent upon and therefore determined by the fourth order moments $E Z_0 Z_m \underline{S}_0^T R_M^{-1} \underline{S}_m$ in the independent case[7]. These moments are, at least for linear processes, determined by the underlying dynamical model (i.e, transfer function) and probability law which govern the process and hence cannot be determined *a priori* without further restrictions on the same. This inescapable fact constitutes the origin of the Minimum FPE conjecture's failure.

As promised at the outset of this section, we now develop several counterexamples to the Minimum FPE conjecture. These counterexamples, as claimed at the close of section 3.1.1, prove conclusively that the conjecture is in fact false.

Counterexamples in the Independent Case

The counterexamples given below demonstrate the remarkable inaccuracy of the FPE statistic as an estimate of the mean square prediction error incurred by the adaptive linear predictor in the independent case[8] (under which the validity of the FPE estimate is claimed). Indeed, the FPE statistic is shown, both with and without the Gaussian assumption, to consistently exhibit a bias in its estimate of the adaptive linear predictor's MSE when the order of the predictor is less than the order of the process to be predicted. This observation provides an explanation of the Minimum FPE procedure's propensity to select an adaptive predictor order that is greater than the actual (or effective) order of the underlying process.

Although the Minimum FPE procedure's tendency to overparameterize the model has been suggested by several authors [18], [22-24], [26-27], [35], [39] on the basis of either empirical data (obtained via simulation) or asymptotic consistency arguments, no closed form proof[9] of the effect has been given. We believe, therefore that the results presented herein represent the first theoretical explanation of this phenomenon and hence proof of the Minimum FPE conjecture's failure.

---

[7] This observation is shown to hold in the dependent case as well, in section 3.2.2.

[8] The FPE statistic is shown, in section 3.2.2, to induce potentially unbounded errors, when employed as an estimate of $\sigma^2[M,N]$ under the more realistic dependent case.

[9] For the finite length realization inherent in practical applications.

This fact becomes apparent upon examination of the following counterexamples.

The first two counterexamples are based on the results of Theorems 3.1 and 3.2 and therefore do not necessitate the Gaussian assumption. The derivation of these examples, however, entails the closed form evaluation of the adaptive linear predictor's MSE from (3.26). Hence, we have chosen rather simple process models and restrict the order of the predictor to unity (i.e., $M=1$), for analytical tractability.

An overview of the lengthy calculations behind the results presented below, as well as proof that the subject counterexamples satisfy the strong mixing requirements of Theorems 3.1 and 3.2, may be found in Appendix 3B.

Counterexample 1

Suppose $\{S_n\}_{n=-\infty}^{+\infty}$ is a 2nd order, zero mean, stationary AR process (i.e., $P=2$ and $\alpha_{P,\infty}=0$ in (2.6)) defined by

$$S_n = \frac{1}{2}S_{n-1} - \frac{1}{2}S_{n-2} + \epsilon_n \ , \quad \{\epsilon_n\}\ i.i.d.\ \ with\ E\,\epsilon_n = 0\ . \tag{3.27}$$

It then follows that this process is strong mixing and satisfies the conditions of Theorems 3.1 and 3.2. Furthermore it can be shown, from (3.17), that

$$\mu(1,N) = \frac{1}{\sigma^2[1,\infty]N}\sum_{m=-N}^{N}(N - |m|)EZ_0Z_m\,\underline{S}_0^T R_M^{-1}\underline{S}_m = \frac{1}{3} + O(\frac{1}{N})\ .$$

Hence the mean square prediction error incurred by a 1st order adaptive linear predictor of this process is, from (3.12),

$$\sigma^2[1,N] = \sigma^2[1,\infty]\left[1 + \frac{1}{3N}\right] + o(\frac{1}{N}) \tag{3.28}$$

and an asymptotically unbiased estimate of this error is, employing (3.26), given by

$$\hat{\sigma}^2[1,N] = F^2(1,N)\left[\frac{1 + \dfrac{1}{3N}}{1 - \dfrac{1}{3N}}\right]. \tag{3.29}$$

The ramifications of (3.28) and (3.29) are discussed following the second counterexample.

Counterexample 2

Let $\{S_n\}_{n=-\infty}^{+\infty}$ be a zero mean, stationary 2nd order moving average (MA) process (i.e., P effectively infinite and $\alpha_{P,\infty}=0$ in (2.6)) specified by

$$S_n = \epsilon_n + \epsilon_{n-1} \quad , \ \{\epsilon_n\}\ \ as\ in\ (3.27)\ . \tag{3.30}$$

This process is clearly strong mixing (as are all m-dependent processes) with a mixing coefficient, α_k, which vanishes for all $k > 1$, and obviously satisfies the requirements of Theorems 3.1 and 3.2.

Applying these theorems we obtain, from (3.17),

$$\mu(1,N) = \frac{2}{3} + O(\frac{1}{N})$$

and, from (3.12), it immediately follows that the 1st order adaptive linear predictor of this process exhibits a MSE given by

$$\sigma^2[1,N] = \sigma^2[1,\infty]\left[1 + \frac{2}{3N}\right] + o(\frac{1}{N}) . \tag{3.31}$$

Moreover, an asymptotically unbiased estimate of this predictor's MSE is determined by (3.26), as

$$\hat{\sigma}^2[1,N] = F^2(1,N)\left[\frac{1 + \frac{2}{3N}}{1 - \frac{2}{3N}}\right] . \tag{3.32}$$

Clearly, the mean square prediction error of the 1st order adaptive predictor in each of the examples above does not agree with the estimate

$$\sigma^2[1,N] = \sigma^2[1,\infty](1 + \frac{1}{N})$$

suggested by the Minimum FPE conjecture. Furthermore, the FPE estimate (from (2.8), as modified to reflect the zero mean assumption in the present case):

$$\hat{FPE}_1 = F^2(1,N)\left[\frac{1 + \frac{1}{N}}{1 - \frac{1}{N}}\right] \tag{3.33}$$

exhibits an unmistakably positive bias relative to the unbiased estimates (3.29) and (3.32) of the adaptive linear predictor's MSE in counterexamples 1 and 2.

Although the bias of the FPE estimate in these counterexamples is, for long learning periods, N, not alarming, it should be noted that the subject examples were chosen only for the relative computational simplicity which they afforded. That these counterexamples in some measure provide a lower bound on the FPE estimate's potential error can be appreciated by examining the magnitude which this estimator's bias can assume. To this end, employing (2.8) and (3.26), we determine the expected bias of the FPE statistic to be

$$\text{Bias } \hat{FPE}_M = E\{\hat{FPE}_M - \hat{\sigma}^2[M,N]\} = 2\sigma^2[M,\infty]\left[\frac{M - \mu(M,N)}{N - M}\right] + o(\frac{M}{N}) \tag{3.34}$$

which, in turn, implies

$$|\text{Bias } \hat{FPE}_M| \geq \frac{2\sigma^2[M,\infty]}{N} |M - \mu(M,N)| + o(\frac{M}{N}) .$$

Observing that $\mu(M,N)$ cannot be bounded (even with M and N fixed) we conclude that it would not require an exhaustive search to locate an example[10] which would expose an inordinately large bias in the FPE estimate. Indeed, we believe that it is entirely possible, if not probable, that such processes will be encountered in typical adaptive linear prediction environments.

At this juncture one might ask why we have not addressed the accuracy of the FPE estimate when the candidate predictor's order, M, is greater than or equal to the order, P, of the process to be predicted (assuming, of course, that the process admits a finite autoregressive representation). This we consider now.

---

[10] Albeit, a computationally difficult one.

Under the assumption that $M \geq P$, the optimal predictor's error sequence $\{Z_n\}$ is identical to the innovations sequence associated with the AR process in question. If it is further assumed that the innovations process is i.i.d., then clearly

$$EZ_0 Z_m \underline{S}_0^T R_M^{-1} \underline{S}_m = \sigma^2[M,\infty] M \delta_{m0}$$

where δ_{m0} is the Kronecker delta function defined by

$$\delta_{m0} = \begin{cases} 1 & m = 0 \\ 0 & m \neq 0 \end{cases} .$$

It then follows from (3.17) and (3.24) that

$$\hat{\sigma}^2[M,N] = F^2(M,N) \left[\frac{1 + \frac{M}{N}}{1 - \frac{M}{N}} \right]$$

as proffered by the Minimum FPE conjecture.

However, the validity of this result, as in its original derivation, is critically dependent upon two premises:

(1) the innovations process underlying the AR process is i.i.d.

and

(2) the independent case assumption - wherein the LMS adaptive predictor, and the observations on which it operates, are derived from statistically independent realizations of the input process.

The innovations process is conventionally taken to be the mean square unpredictable component of the AR process. The i.i.d. assumption represents a stronger condition[11] (unless, of course, the AR process is Gaussian) and thus undermines the credibility of Akaike's result. We defer discussion on the limitations of the second assumption to section 3.2.1, except to note that it is widely considered unrepresentative of and therefore restrictive in typical adaptive prediction environments.

The fact that the FPE statistic constitutes an accurate estimate, albeit under implausible premises, of an Mth order adaptive predictor's MSE when applied to an Pth order AR process if $M \geq P$, does not infer that the conjecture is valid. On the contrary, the demonstrated inaccuracy of this statistic, when $M < P$, can unduly bias and thus detrimentally dominate the order selection decision under the Minimum FPE conjecture.

Clearly then, the Minimum FPE procedure must be considered unreliable unless *a priori* knowledge of an upper bound on the process order is available. In this unlikely event, the procedure would lead to the selection of the minimum candidate predictor order provided that all admissible orders were in fact restricted to be greater than or equal to the aforementioned *a priori* bound. This order estimate would however, by construction, exhibit a positive bias. Thus it seems that bias is an inescapable product of the Minimum FPE procedure.

Such potentially excessive biases induce substantial inaccuracies in the FPE order selection procedure thereby rendering the conjecture unreliable and therefore invalid. In further support of this conclusion, we demonstrate below that, if a

[11] Hannan [39] maintains that resorting to such a device constitutes "... a fiction.".

slightly more restrictive hypothesis is adopted, it becomes quite easy to identify and evaluate counterexamples that induce results which deviate significantly from those suggested by the Minimum FPE conjecture.

Theorem 3.3

Let $\{S_n\}_{n=-\infty}^{+\infty}$ be a zero mean, stationary Gaussian process with an unknown power spectral density $P_{SS}(e^{i2\pi\lambda}) \in L_2[-\frac{1}{2}, \frac{1}{2}]$.[12]

Then the mean square prediction error of the adaptive linear predictor of this process is, in the *independent case*, given by

$$\sigma^2[M,N] = \sigma^2[M,\infty]\left[1 + \frac{\mu(M)}{N}\right] + o(\frac{M}{N}) \tag{3.35}$$

with

$$\mu(M) = \frac{1}{2\pi i}\oint_C \left[G(z)G(z^{-1})\underline{z}^T R_M^{-1}\underline{z}^{-1} + G^2(z)\underline{z}^T R_M^{-1}\underline{z}\right]P_{SS}^2(z)\frac{dz}{z}(\sigma^2[M,\infty])^{-1} \tag{3.36}$$

where

$\underline{z} = (z, z^2, \cdots, z^M)^T$, $\underline{z}^{-1} = (z^{-1}, z^{-2}, \cdots, z^{-M})^T$, and $G(z) = 1 - \underline{\beta}_{M,\infty}^T\underline{z}^{-1}$.

(Here we employ z, as is customary, to denote the general variable of the complex plane and hope that confusion with the error sequence Z_n will not arise.)

Equation (3.35) defines the mean square error of an Mth order LMS adaptive predictor, under the assumption that the predictor is derived from, and applied to, independent realizations of a zero mean Gaussian process determined by a square integrable psd $P_{SS}(z)$. However, the MSE is specified in terms of the Mth order optimal predictor and psd of the process in question and is thus of limited utility in adaptive prediction applications. Nonetheless, Theorem 3.3 establishes an accurate reference against which the validity of the Minimum FPE conjecture and its associated statistic may be evaluated.

With the benefit of Davisson's work [8], the following result is an immediate consequence of Theorem 3.3.

Corollary 3.1

Let $\{S_n\}_{n=-\infty}^{+\infty}$ satisfy the hypothesis of Theorem 3.3.

Then an asymptotically unbiased estimator of the adaptive linear predictor's MSE in the *independent case* is realized by

$$\hat{\sigma}^2[M,N] = F^2(M,N)\left[\frac{1 + \frac{\mu(M)}{N}}{1 - \frac{\mu(M)}{N}}\right] \tag{3.49}$$

where $\mu(M)$ is defined by (3.36).

---

[12] This hypothesis is due to Davisson [7,8] who employed it to determine the adaptive predictor's MSE under the dependent case assumption.

Furthermore, if the process is also P th order autoregressive (i.e., P th order Gauss-Markov) and $M \geq P$, then $\mu(M) = M$ and

$$\hat{\sigma}^2[M,N] = F^2(M,N) \left[\frac{1 + \frac{M}{N}}{1 - \frac{M}{N}} \right] . \tag{3.50}$$

The independent case estimates of the adaptive predictor's MSE given by (3.49) and (3.50) are clearly defined in terms, and therefore assume knowledge, of specific characteristics of the process to be predicted. The first estimate (3.47) requires a complete specification of both the processes' psd and the coefficients of its Mth order optimal predictor. The second estimate is somewhat less demanding, requiring "only" an *a priori* assurance that the process in question is in fact Gauss-Markov and an upper bound on the regression order. This information is conspicuously absent in typical adaptive prediction applications; consequently these estimates have little if any practical value in such environments.

However, the estimates in question were neither derived nor intended for such use. They were formulated solely to provide an absolute reference for measuring the accuracy of the FPE statistic and consequently, the validity of the conjecture as well. The results of this analysis follow.

Comparing (2.7) and (2.8) to (3.35) and (3.49), respectively, we see that the error factor $\mu(M)$ in the latter equations corresponds to the factor M (i.e., assuming a zero mean process) in equations (2.7) and (2.8) of Akaike's Minimum FPE conjecture. Thus the validity of Akaike's equations, and hence of the Minimum FPE conjecture, for Gaussian processes, hinges on the behavior of $\mu(M)$ as a function of M.

This observation can, perhaps, be more readily appreciated when examining the bias induced by the FPE estimate of the adaptive predictor's MSE. Thus, from (2.8) and (3.49) with (3.51) we find that

$$\text{Bias}\, \hat{FPE}_M = E\{\hat{FPE}_M - \hat{\sigma}^2[M,N]\} = 2\sigma^2[M,\infty] \left[\frac{M - \mu(M)}{N - M} \right] + o\left(\frac{M}{N}\right) \tag{3.57}$$

which implies

$$|\text{Bias}\, \hat{FPE}_M| \geq \frac{2\sigma^2[M,\infty]}{N} |M - \mu(M)| + o\left(\frac{M}{N}\right) .$$

Thus we see that the bias of the FPE statistic is clearly determined by the behavior of $\mu(M)$.

An examination of (3.36) reveals that if the process in question were indeed white (i.e., i.i.d., since Theorem 3.3 assumes the Gaussian hypothesis) then $G(z) = 1$, and $P_{SS}(z) = \sigma^2[M,\infty]$ for all $M \geq 1$, which implies $\mu(M) = M$ for all $M \geq 1$. Moreover, since the error factor $\mu(M)$ is a continuous functional in $G(z)$ and $P_{SS}(z)$, this factor approaches M as the process to be predicted becomes mean square unpredictable. This observation clearly exposes the irony of the FPE procedure - the error factor, M, conjectured by the FPE statistic, is found to be in agreement with the actual error factor, $\mu(M)$, when the theory becomes useless.

Further consideration of (3.36) suggests that the behavior of the error factor is neither obvious nor readily computable for any non-trivial psd. Therefore, an APL computer program was written to compute $\mu(M)$, in closed form, by applying the Cauchy Residue Theorem to evaluate the complex contour integral contained in (3.36). The program's accuracy was validated when its outputs were found to be in agreement with results obtained analytically in counterexamples 1 and 2. This program was subsequently used to determine the behavior of the error factor $\mu(M)$, as a function of the adaptive predictor's order, M, for each of several AR processes (i.e., AR processes of order $P = 2$ through 16, with varying psd's).

The error factor, $\mu(M)$, exhibited a consistent behavior, as a function of M and process psd, over all process models tested. Figures 1 and 2 depict plots of $\mu(M)$, for 5th and 7th order AR processes, respectively, which demonstrate the typical functional dependence of $\mu(M)$ on the processes' psd and the adaptive predictor's order M. The solid line in each plot represents the error factor $\mu(M) = M$, as suggested by the Minimum FPE conjecture; i.e., as independent of the processes' psd. The symbols plotted in the figures reflect the actual values of $\mu(M)$ computed by the APL program as a function of both M and the psd of the process in question. Each unique symbol indicates the psd, shown on the right-hand side of the figures, of the process for which the corresponding values of $\mu(M)$ were computed.

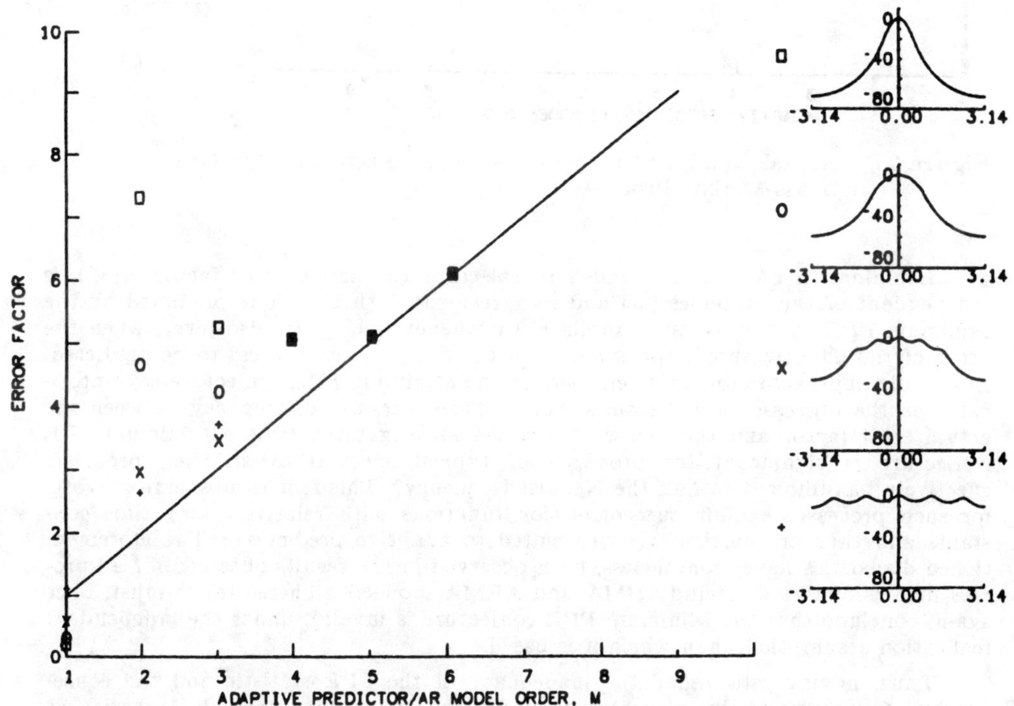

Figure 1 Actual, $\mu(M)$, and Predicted, M, Error Factors for 5th Order Gauss-Markov Processes

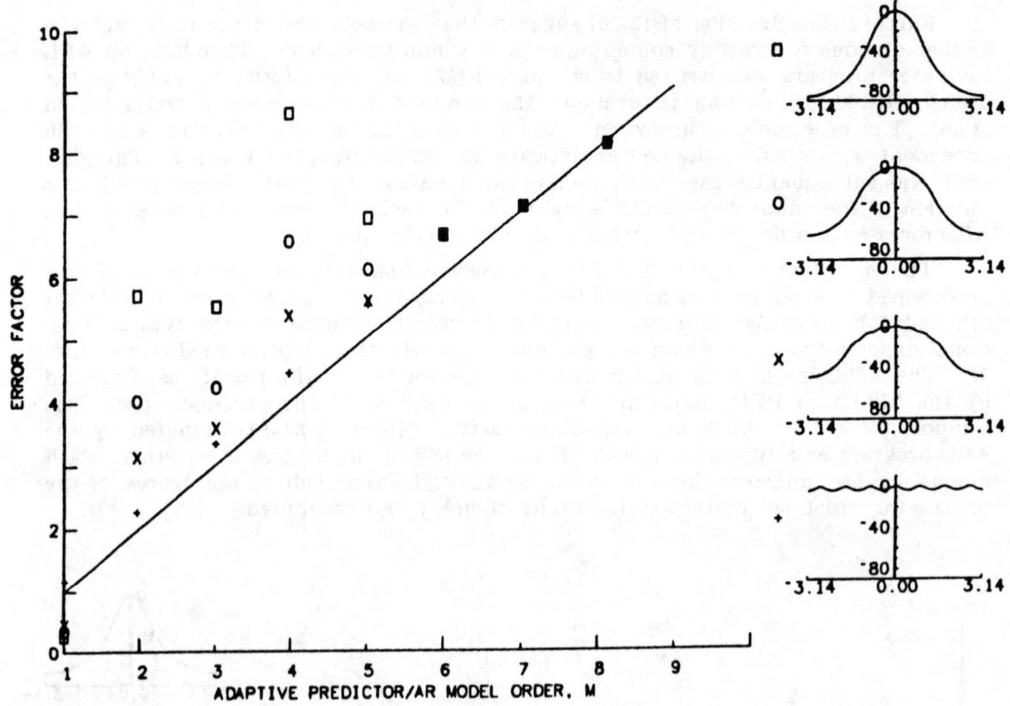

Figure 2 Actual, $\mu(M)$, and Predicted, M, Error Factors for 7th Order Gauss-Markov Processes

Examination of Figures 1 and 2 reveals that the actual error factor, $\mu(M)$, is independent of the processes psd and in agreement with the value predicted by the Minimum FPE conjecture and Corollary 3.1 whenever $M \geq P$. However, when the order of the adaptive predictor is less than the order of the process to be predicted, $\mu(M)$ only approaches the value claimed by the Minimum FPE conjecture asymptotically, as the processes' psd becomes white. Moreover, the discrepancy between the actual error factor and the factor M erroneously suggested by the Minimum FPE conjecture is significant for processes of typical spectral extent; i.e., processes effectively bandlimited to half the Nyquist frequency. This fact is most unfortunate, for such processes exhibit autocorrelation functions with relatively long time constants and thus are particularly well suited to adaptive prediction. The aforementioned disparities have, nonetheless, been observed in the results obtained for all process models analyzed (including MA and ARMA models). Therefore we must, once again, conclude that the Minimum FPE conjecture is invalid, under the independent realization assumption upon which it is based.

Thus, having established the inadequacy of the FPE statistic and the consequential failure of Akaike's conjecture under the independent case hypothesis, we conclude our study of the adaptive linear predictor, in this case, with the derivation of a frequency domain formula for its MSE. This result is subsequently compared with and verified by (or verifies) the findings of Kunitomo and Yamamoto [31], for completeness.

3.1.3 The Adaptive Linear Predictor's MSE in the Independent Case

In this section we develop a frequency domain based formula for the mean square prediction error incurred by an adaptive linear predictor when applied to a stationary linear process in the independent case. This result requires, in addition to our usual assumption of regularity (cf. Chapter 1 section 1.1 and Appendix 1A), only that the associated innovations process is uncorrelated through fourth order moments. These assumptions clearly represent a relaxation of both the i.i.d. innovations assumption invoked by many authors, and the conditions prerequisite to our earlier results; i.e., the innovations sequence underlying regular (and therefore linear) strong mixing or Gaussian processes exhibits moments which are uncorrelated through fourth order. Hence the results obtained in the preceding section become a special case of the theory presented herein.

Before proceeding with the development of the main result, we make the following observation[13] which proves to be quite useful in the sequel.

Observation:

Suppose $\{S_n\}_{n=-\infty}^{+\infty}$ is a zero mean, stationary, regular discrete time parameter stochastic process with an absolutely continuous spectral distribution.
Then $\{S_n\}$ can be represented as the output of a linear time invariant system

$$S_n = \sum_{k=-\infty}^{+\infty} h_k W_{n-k} \quad , \quad EW_n = 0 \quad , \quad EW_k W_l = \sigma_W^2 \delta_{kl}$$

and admits the spectral representation

$$S_n = \frac{1}{2\pi} \int_{-\pi}^{\pi} S(\lambda) e^{in\lambda} d\lambda \tag{3.58}$$

where $S(\lambda) = H(\lambda) W'(\lambda)$ with

$$H(\lambda) = \sum_{k=-\infty}^{+\infty} h_k e^{-ik\lambda} \quad , \quad W(\lambda) = \sum_{k=-\infty}^{+\infty} W_k \{\frac{1-e^{-ik\lambda}}{ik}\}$$

and all equalities involving infinite sums are taken to be in the mean; i.e.,

$$S(\lambda) = \sum_{k=-\infty}^{+\infty} S_k e^{-ik\lambda} \iff S(\lambda) = \underset{N\to\infty}{\text{l.i.m.}} \sum_{k=-N}^{N} S_k e^{-ik\lambda} .$$

Theorem 3.4

Let $\{S_n\}_{n=-\infty}^{+\infty}$ be a zero mean, stationary, regular discrete time parameter stochastic process with an absolutely continuous spectral distribution.
Assume further that the stationary process $\{W_n\}_{n=-\infty}^{+\infty}$, as defined in (3.58), is uncorrelated through fourth moments; i.e.,

$$EW_{n_1} W_{n_2} W_{n_3} W_{n_4} = \begin{cases} EW_{n_i}^2 W_{n_j}^2 = \sigma_W^4 & n_i = n_k \ , \ n_j = n_l \ , \ n_i \neq n_j \ , \ 1 \le i,j,k,l \le 4 \\ EW_0^4 & n_1 = n_2 = n_3 = n_4 \\ 0 & else \end{cases} . \tag{3.59}$$

Then the mean square prediction error manifested by the Mth order adaptive linear predictor of the process $\{S_n\}$ is given, in the *independent case*, by

[13] cf. Doob [40] pp. 481-483 and 499-500 for the theory behind and a proof of the observation.

$$\sigma^2[M,N] = \sigma^2[M,\infty]\left[1 + \frac{\mu(M)}{N}\right] + o\left(\frac{M}{N}\right) \tag{3.60}$$

where

$$\mu(M) = \frac{\sigma_W^2 \int_{-\pi}^{\pi} |H(\lambda)|^4 \left[|\Phi(\lambda)|^2 \underline{T}^T(\lambda) R_M^{-1} \underline{T}^*(\lambda) + (\Phi^*(\lambda))^2 \underline{T}^T(\lambda) R_M^{-1} \underline{T}(\lambda) \right] d\lambda}{\int_{-\pi}^{\pi} |H(\lambda)|^2 |\Phi(\lambda)|^2 d\lambda}$$

and

$$H(\lambda) = \sum_{k=-\infty}^{+\infty} h_k e^{-ik\lambda} \quad , \quad \Phi(\lambda) = 1 - \underline{\beta}_{M,\infty}^T \underline{T}(\lambda) \quad , \quad \underline{T}(\lambda) = (e^{-i\lambda}, e^{-i2\lambda}, \cdots, e^{-iM\lambda})^T .$$

It is clear, upon reviewing (3.60), that the spectral formula for the error factor $\mu(M)$ is not readily evaluated. However, a comparison of (3.60) with the analogous result given, under the Gaussian hypothesis, by (3.36) of Theorem 3.3 reveals that these formulas are functionally identical. To confirm this observation, one need only evaluate $\mu(M)$, as specified by (3.36), on the unit circle $z = e^{i\lambda}$, $\lambda \in [-\pi, \pi]$.

It therefore follows that the analysis of and the conclusions drawn on the behavior of $\mu(M)$ following Theorem 3.3 are also applicable here. This of course lends further support to our earlier finding that the Minimum FPE conjecture cannot be substantiated in the independent case under which it is postulated.

3.1.4 Comparison with the Findings of Kunitomo and Yamamoto

As noted in Chapter 1, the most comprehensive (prior) research into the performance of adaptive linear predictors is due to Kunitomo and Yamamoto [31]. Recognizing this, we believe that a comparison of these findings, with the results obtained herein, is not only appropriate, but is also essential for completeness.

Kunitomo and Yamamoto derive estimates of the LMS adaptive linear predictor's MSE, through terms of $O(N^{-1})$, under both the independent and dependent cases, based upon the assumption that the process to be predicted is Pth order AR[14]. The results of these derivations, as stated in Corollary 2 and Theorem 3 of the cited work, although elegant, are defined in terms of infinite sums of the autocorrelation function and autoregression coefficients of the process to be predicted. The corresponding results of our research, as presented in Theorems 3.1, 3.3, and 3.4 of sections 3.1.2 and 3.1.3, are not predicated on an autoregressive process model. Thus a direct comparison of the formulae in question would be difficult at best.

In view of these difficulties, we resorted to employing Kunitomo and Yamamoto's formulae to evaluate the MSE of the adaptive predictor's in the specific counterexamples 1 and 2, as defined in section 3.1.2. The results of these computations (after inverting the 2nd order MA model of the second counterexample to obtain its equivalent infinite AR representation) proved to be in perfect agreement with the results given in (3.28) and (3.31). We therefore conclude that it is highly probable that our theory and the results of Kunitomo and Yamamoto, despite cursory appearances to the contrary, agree in the general case as well.

[14] Subject to additional constraints as outlined in Chapter 1.

3.2 Considerations in Extending Davisson's Method

In this section we begin the process of addressing the issues surrounding and impediments to achieving our stated objective of extending Davisson's results and Method beyond the limitations imposed by the Gaussian assumption. The first issue considered involves the question of the legitimacy of the independent case hypothesis, and the reliability of results derived under this assumption, in adaptive prediction environments. Thus, the consequences of the independent case hypothesis are reviewed and this assumption is in fact shown to be a device of analytic convenience only - the adoption of which is attended by the inevitable loss of accuracy and applicability that usually accompanies such model oversimplifications. We conclude, as have other rigorous authors, that any result which is to accurately reflect the conditions and performance expected in typical adaptive linear prediction applications must be based on the realistic, but analytically difficult, dependent case hypothesis.

In view of this conclusion, we reexamine the mean square prediction error incurred by the adaptive predictor under the realistic dependent case model. The results of these analyses are presented in two theorems which reveal that, under this hypothesis, the adaptive linear predictor's MSE is dependent upon the ratio of the fourth and the square of the second order moments of the innovations process associated with the process to be predicted. This dependence, as shown by revisiting the counterexamples of section 3.1.2, is not reflected by and constitutes the primary source of considerable error in estimates of the adaptive predictor's MSE derived under the independent case assumption. Consequently, order selection criteria based on such independent case estimates suffer from a lack of accuracy. This observation provides the final proof necessary to justify the dismissal of the Minimum FPE conjecture as a viable candidate for the extension of Davisson's Method.

Davisson's results are revisited and found, unfortunately, to be biased under the restrictions inherent in finite model order and observation sequence length applications. Although this shortcoming was originally recognized by Davisson, it nonetheless constitutes yet another obstacle that must be overcome in extending and enhancing his results.

3.2.1 The Inadequacy of the Independent Case Assumption

The independent case assumption is invoked by most investigators of adaptive linear prediction (e.g., Akaike, *et. al.*) as an expedient means of significantly reducing the analytical difficulty encountered in characterizing the MSE performance of the inherently nonlinear, stochastic, adaptive predictor. The independent case hypothesis ensures (at least in the investigator's mind) the statistical decoupling of the adaptive predictor's random coefficient estimation errors from the data to which the adapted coefficients are applied - thereby rendering the problem tractable.

Unfortunately, this oversimplification also guarantees a commensurate loss of accuracy in the model thus obtained.[15] This observation is confirmed in section 3.2.2, where it is shown that estimates of the adaptive linear predictor's MSE, obtained under the independent case device (e.g., the FPE statistic), deviate dramatically from from the corresponding estimates derived under the more realistic dependent case hypothesis. Our conclusion is supported by the work [31] of Kunitomo and Yamamoto who note that "...the differences between the dependent and independent case [results] can be substantial .... This raises some question about the conventional [independent case] assumption .".

[15] This consequence is, or should be, a fundamental axiom of scientific investigation.

The significance of these disparities is amplified when one recognizes that the dependent case hypothesis reflects the conditions most likely to exist in adaptive prediction applications. Indeed, in such environments, only a single finite length realization of the process to be predicted is observable. Moreover, the processes in question are often effectively bandlimited and thus manifest autocorrelation functions which decay slowly relative to the length $(N+M)$ of the observation sequence. In such applications, the adaptive predictor's estimated coefficients are statistically dependent upon the data to which they are applied thereby rendering the independent case assumption unjustifiable.

Furthermore, in those rare situations where a second, statistically identical, but independent observation sequence is available, one must ask why this additional information would not be used by the adaptation mechanism in an attempt to improve its estimate of the optimal prediction coefficients. To fail to do so, in the stationary environment implied, clearly violates one of the most fundamental principles of both estimation and information theory.

Thus, in view of these observations, we must conclude that the independent case hypothesis, as well as any estimate or method based on this assumption, is unacceptable. Accordingly, after presenting the analysis that supports the claims made above, we restrict our attention to the dependent case hypothesis for the remainder of this treatise.

3.2.2 The Adaptive Linear Predictor's MSE in the Dependent Case

We now examine the mean square prediction error incurred by the Mth order adaptive linear predictor in the *dependent case.* Having established the adaptive predictor's MSE under this hypothesis, we derive an estimate of the error and uncover the difficulties in quantifying the estimate thus obtained. These results are developed in a series of theorems which constitute the *dependent case* analogues of the theorems given in sections 3.1.2 and 3.1.3 under the *independent case* hypothesis.

Theorem 3.5

Let $\{S_n\}_{n=-\infty}^{+\infty}$ be a zero mean, stationary strong mixing process with mixing coefficient α_k such that

$$\sum_{k=1}^{\infty} \alpha_k^{\frac{\delta}{2+\delta}} < \infty \quad \text{and} \quad E\,|S_n|^{8(2+\delta)} < \infty \quad \textit{for some} \quad \delta > 0\,. \tag{3.70}$$

Then the mean square prediction error, $\sigma^2[M,N]$, of the Mth order adaptive linear predictor is given, in the *dependent* case, by

$$\sigma^2[M,N] = \sigma^2[M,\infty] + \frac{1}{N} E Z_n^{\,2} \underline{S}_n^T R_M^{-1} \underline{S}_n + o\left(\frac{M}{N}\right) \tag{3.71}$$

where, as usual, $Z_n = S_n - \underline{\beta}_{M,\infty}^T \underline{S}_n$ denotes the error incurred by the Mth order optimal predictor.

The proof of Theorem 3.5 is quite lengthy and thus has been relegated to Appendix 3D. This proof nonetheless warrants consideration for it reveals the rather elegant manner in which the strong mixing hypothesis ensures the given result.

Before proceeding with a discussion of the implications of Theorem 3.5, we define the *excess* MSE of the adaptive linear predictor. This term will be used in the analyses which follow.

Definition:

The *excess* MSE of the Mth order adaptive linear predictor is that component of its mean square prediction error which is in excess of the MSE exhibited by the optimal predictor of the same order; i.e.,

$$\text{excess MSE} = \sigma^2[M,N] - \sigma^2[M,\infty] .$$

Intuitively, the excess MSE is both a consequence of and proportional to the inherent error incurred by the adaptive predictor in estimating the coefficients of the optimal predictor, on the basis of a finite number (N) of observations.

Comparing (3.71) with the analogous result (3.12) for the independent case, one arrives at the unexpected conclusion that the functional form of the adaptive predictor's excess MSE under the dependent case hypothesis is a special case of the form assumed by the predictor's excess error under the independent case assumption. This initial impression is however somewhat misleading. Indeed, we demonstrate below that, in the dependent case, the adaptive predictor's excess MSE, as normalized by the optimal predictor's MSE, is explicitly dependent upon the second and fourth moments of the innovations process. This represents a remarkable deviation from the corresponding finding in the independent case - where the ratio of the adaptive predictor's excess and the optimal predictor's mean square errors were shown to be independent of the probability law governing the innovations process (cf. section 3.1.2, counterexamples 1 and 2).

Nevertheless some similarities between the two cases remain. In particular, we note, from (3.71), that if the optimal predictor's MSE as well as the fourth order moments $EZ_n^2\underline{S}_n^T R_M^{-1}\underline{S}_n$ were known (improbable in adaptive prediction applications) then an asymptotically unbiased estimate of the mean square prediction error incurred by the adaptive linear predictor in the dependent case would be defined by

$$\hat{\sigma}^2[M,N] = \sigma^2[M,\infty]\left[1 + \frac{\mu_D(M)}{N}\right] \tag{3.72}$$

where[16] $\mu_D(M) = (\sigma^2[M,\infty])^{-1} EZ_n^2\underline{S}_n^T R_M^{-1}\underline{S}_n$.

Moreover, the results of Theorem 3.2 remain valid under the dependent case hypothesis. Therefore, combining (3.72) and (3.24) we observe that an asymptotically unbiased estimate of the adaptive predictor's MSE under the dependent case is also realized by

$$\hat{\sigma}^2[M,N] = F^2(M,N)\left[\frac{1 + \frac{\mu_D(M)}{N}}{1 - \frac{\mu(M,N)}{N}}\right] \tag{3.73}$$

with $\mu(M,N)$ as defined in (3.17).

As noted during the course of the independent case analysis, these estimates require knowledge of the aforementioned fourth moments as well as the optimal predictor's MSE and consequently are of theoretical interest only.

With the results derived from Theorem 3.5, we are in a position to reexamine the counterexamples, of the preceding section, under the dependent case hypothesis.

---

[16] The subscript D is adopted to distinguish the dependent case result.

Counterexamples in the Dependent Case

We revisit the counterexamples of section 3.1.2 to prove that the error inherent in the FPE estimate of the adaptive linear predictor's MSE cannot be bounded under the realistic dependent case hypothesis. The analysis and calculations supporting the results presented below are provided in Appendix 3B. The counterexample process model definitions are repeated here for the convenience of the reader.

Counterexample 1

Suppose $\{S_n\}_{n=-\infty}^{+\infty}$ is a 2nd order, zero mean, stationary AR process (i.e., $P = 2$ and $\alpha_{P,\infty} = 0$ in (2.6)) defined by

$$S_n = \frac{1}{2}S_{n-1} - \frac{1}{2}S_{n-2} + \epsilon_n \quad , \quad \{\epsilon_n\} \; i.i.d. \; with \; E\,\epsilon_n = 0 \, . \tag{3.74}$$

It then follows, from (3.72), that

$$\mu_D(1) = 0.88 + 0.04\frac{E\,\epsilon_n^4}{(E\,\epsilon_n^2)^2} \tag{3.75}$$

and the mean square prediction error of the 1st order adaptive linear predictor of this process in the dependent case is, from (3.71), found to be

$$\sigma^2[1,N] = \sigma^2[1,\infty]\left[1 + \frac{1}{N}\left[0.88 + 0.04\frac{E\,\epsilon_n^4}{(E\,\epsilon_n^2)^2}\right]\right] + o(\frac{1}{N}) \, . \tag{3.76}$$

Moreover, an asymptotically unbiased estimate of this predictor's MSE in the dependent case is, from (3.29) and (3.73),

$$\hat{\sigma}^2[1,N] = F^2(1,N)\left[\frac{1 + \frac{1}{N}\left[0.88 + 0.04\frac{E\,\epsilon_n^4}{(E\,\epsilon_n^2)^2}\right]}{1 - \frac{1}{3N}}\right] . \tag{3.77}$$

The implications of these findings are discussed following the second counterexample.

Counterexample 2

Let $\{S_n\}_{n=-\infty}^{+\infty}$ be a zero mean, stationary 2nd order moving average (MA) process (i.e., P effectively infinite and $\alpha_{P,\infty} = 0$ in (2.6)) specified by

$$S_n = \epsilon_n + \epsilon_{n-1} \quad , \quad \{\epsilon_n\} \; as \; in \; (3.27) \, . \tag{3.78}$$

Then, from (3.72), we find that

$$\mu_D(1) = \frac{1}{2} + \frac{1}{6}\frac{E\,\epsilon_n^4}{(E\,\epsilon_n^2)^2} \tag{3.79}$$

and that the 1st order adaptive linear predictor of this process manifests a MSE in the dependent case, as determined from (3.71), given by

$$\sigma^2[1,N] = \sigma^2[1,\infty]\left[1 + \frac{1}{N}\left[\frac{1}{2} + \frac{1}{6}\frac{E\,\epsilon_n^4}{(E\,\epsilon_n^2)^2}\right]\right] + o(\frac{1}{N}) \, . \tag{3.80}$$

Furthermore, an asymptotically unbiased estimate of this predictor's error is found in the dependent case, from (3.32) and (3.73), to be

$$\hat{\sigma}^2[1,N] = F^2(1,N)\left[\frac{1 + \frac{1}{N}\left[\frac{1}{2} + \frac{1}{6}\frac{E\epsilon_n^4}{(E\epsilon_n^2)^2}\right]}{1 - \frac{2}{3N}}\right]. \qquad (3.81)$$

The results, (3.76), (3.77), (3.80) and (3.81), thus derived under the dependent case hypothesis should be compared with the corresponding findings, (3.28), (3.29), (3.31), and (3.32), respectively, obtained under the independent case assumption. In doing so we find that the MSE exhibited by the adaptive linear predictor as well as the asymptotically unbiased estimates of this parameter differ substantially under the two hypotheses. Indeed, the independent case based results would indicate that the predictor's MSE, and hence any unbiased estimate of this performance measure, are independent of the probability distribution which defines the innovations process underlying the process to be predicted.

This, unfortunately, is not the case. As seen in the more realistic dependent case counterexample results, and demonstrated to hold for the general dependent case in the sequel, the adaptive predictor's excess MSE is in fact a linear function of the *coefficient of kurtosis* (i.e., the ratio of the fourth central moment and the square of the variance) of the innovations process. Hence the adaptive predictor's MSE depends explicitly, through this coefficient, on the probability distribution of the innovations process.

All that can be inferred with respect to the magnitude of the coefficient of kurtosis, without prior knowledge of the probability law governing the innovations process, is

$$\frac{E\epsilon_n^4}{(E\epsilon_n^2)^2} \geq 1 \; . \qquad (3.82)$$

Thus, the aforementioned discrepancy in the independent case based estimates may be arbitrarily large. We conclude, therefore, that statistics derived under and methods based on the independent case assumption (e.g., the FPE statistic and conjecture) can exhibit unbounded error when applied in typical adaptive prediction environments. Consequently, the independent case assumption is dismissed from further consideration, in spite of the analytical simplifications which it affords.

Returning to the dependent case analysis, we note that if the innovations processes in the subject counterexamples are Gaussian, then the MSE effected by the 1st order adaptive linear predictors of the (Gaussian) processes (3.74) and (3.78) becomes, from (3.76) and (3.80),

$$\sigma^2[1,N] = \sigma^2[1,\infty]\left[1 + \frac{1}{N}\right] + o\left(\frac{1}{N}\right)$$

as found in the original theory of Davisson. Moreover, this agreement extends to the general Gaussian dependent case as well. The latter observation is confirmed by recalling that the optimal predictor's error Z_n and the data $\underline{S}_n$, on which the prediction is based, are uncorrelated, and, since these random variates are Gaussian, they are statistically independent as well.

It then immediately follows that any measurable function of the predictor's error is independent of any measurable function of the observations which gave rise to this error. Hence, from (3.71) of Theorem 3.5, we have

$$EZ_n^2 \underline{S}_n^T R_M^{-1} \underline{S}_n = EZ_n^2 E \underline{S}_n^T R_M^{-1} \underline{S}_n = \sigma^2[M,\infty]M$$

which implies

$$\sigma^2[M,N] = \sigma^2[M,\infty]\left[1 + \frac{M}{N}\right] + o\left(\frac{M}{N}\right) \tag{3.83}$$

as formulated by Davisson in 1965.

Regrettably, this agreement does not extend to Davisson's finite length realization approximation of an asymptotically unbiased estimate of the adaptive linear predictor's MSE given (from (2.5)) by

$$\hat{\sigma}^2[M,N] = F^2(M,N)\left[\frac{1+\dfrac{M}{N}}{1-\dfrac{M}{N}}\right]. \tag{3.84}$$

In particular, a discrepancy exists between the denominator of (3.84) (with $M = 1$) and the denominators of the asymptotically unbiased estimates of the predictors' mean square errors, as calculated in (3.77) and (3.81) for the counterexamples above. Moreover, with (3.83) and Theorems 3.2 and 3.3, we find that an asymptotically unbiased estimate of the mean square prediction error exhibited by an Mth order adaptive predictor of a stationary ergodic Gaussian process is given by

$$\hat{\sigma}_*^2[M,N] = F^2(M,N)\left[\frac{1+\dfrac{M}{N}}{1-\dfrac{\mu(M)}{N}}\right] \tag{3.85}$$

where the definition of $\mu(M)$, as given in (3.36), remains valid in the present dependent case.

In view of (3.85), it is clear that the aforementioned discrepancy arises from and is dictated by the difference

$$|M - \mu(M)|$$

in the general Gaussian dependent case as well (cf. comments on the behavior of $\mu(M)$ following Corollary 3.1). It then follows, from (3.85), that the expected bias in the estimate (3.84) of the Mth order adaptive linear predictor's MSE is in the present case given by

$$\text{Bias}\ \hat{\sigma}^2[M,N] = E\{\hat{\sigma}^2[M,N] - \hat{\sigma}_*^2[M,N]\} \tag{3.86}$$

$$= \sigma^2[M,\infty]\left[\frac{(N+M)(M-\mu(M))}{N(N-M)}\right] + o\left(\frac{M}{N}\right)$$

which, asymptotically ($N >> M$), implies

$$|\text{Bias}\ \hat{\sigma}^2[M,N]| \approx \frac{\sigma^2[M,\infty]}{N}\,|M - \mu(M)|\,.$$

Although this bias is not excessive, it does demonstrate the difficulty in attempting to either estimate or bound the fourth order moments found in (3.18) and (3.71) and thus is cause for concern. Indeed, this unfortunate fact was recognized by Davisson [8] in 1966, and led to his disillusionment with the method which now bears his name.

We conclude this section with the derivation of a spectral formula for the mean square error of the adaptive linear predictor under the dependent case. This result, although analogous to the formulation found in Theorem 3.4 of the preceding section, does not rely on the contrived independent case assumption and hence accurately reflects the MSE performance of the predictor in typical adaptive prediction environments. However, the departure from the independent case hypothesis has its price - the theorem obtained by doing so, as will be seen, is of limited practical utility. A study of the theorem's implications nonetheless provides valuable insight into the relationship between the adaptive predictor's error and the psd of, and the probability law governing the innovations sequence associated with, the process to be predicted.

Theorem 3.6

Suppose $\{S_n\}_{n=-\infty}^{+\infty}$ is a zero mean, stationary, regular discrete time parameter stochastic process with an absolutely continuous spectral distribution.

Assume further that the stationary innovations process $\{W_n\}_{n=-\infty}^{+\infty}$ underlying the linear process $\{S_n\}$, as defined in (3.58), is uncorrelated through fourth moments, as specified by (3.59).

Then the mean square prediction error induced by the Mth order adaptive linear predictor of the process $\{S_n\}$ is given, in the *dependent case*, by

$$\sigma^2[M,N] = \sigma^2[M,\infty]\left[1 + \frac{\mu_D(M)}{N}\right] + o\left(\frac{M}{N}\right) \tag{3.87}$$

where

$$\mu_D(M) = M + \frac{\dfrac{\kappa_4}{(2\pi)^3}\displaystyle\int_{-\pi}^{\pi}\int_{-\pi}^{\pi}\int_{-\pi}^{\pi} H(\lambda)H(\mu)H(\nu)H^*(\lambda+\mu+\nu)\Phi(\lambda)\Phi^*(\lambda+\mu+\nu)\underline{T}^T(\nu)R_M^{-1}\underline{T}(\mu)\,d\lambda\, d\mu\, d\nu}{\dfrac{\sigma_W^2}{2\pi}\displaystyle\int_{-\pi}^{\pi}|H(\lambda)|^2|\Phi(\lambda)|^2 d\lambda}$$

with $\kappa_4 = EW_n^4 - 3\sigma_W^4$ and

$$H(\lambda) = \sum_{k=-\infty}^{+\infty} h_k e^{-ik\lambda}, \quad \Phi(\lambda) = 1-\underline{\beta}_{M,\infty}^T\underline{T}(\lambda), \quad \underline{T}(\lambda) = (e^{-i\lambda}, e^{-i2\lambda}, \cdots, e^{-iM\lambda})^T .$$

Although Theorem 3.6 does not, upon initial examination, appear to confirm the results of and observations on the preceding counterexamples, further inspection reveals that the theorem does in fact substantiate these earlier findings. This confirmation can be realized by rearranging the excess MSE factor $\mu_D(M)$.

To this end, recall that the klth element of R_M is given by

$$\{R_M\}_{k,l} = \frac{\sigma_W^2}{2\pi}\int_{-\pi}^{\pi}|H(\lambda)|^2 e^{i(k-l)\lambda}d\lambda \quad , \quad k,l = 1,\cdots,M .$$

Hence we may write $R_M = \sigma_W^2 A_M$, where A_M is the obvious, positive definite matrix. With this factorization, the excess MSE factor $\mu_D(M)$, from (3.87), becomes

$$\mu_D(M) = M + C(H)\left[\frac{EW_n^4}{\sigma_W^4} - 3\right] \tag{3.93}$$

where $C(H)$ is the (constant) factor

$$C(H) = \frac{\frac{1}{(2\pi)^3}\int_{-\pi}^{\pi}\int_{-\pi}^{\pi}\int_{-\pi}^{\pi} H(\lambda)H(\mu)H(\nu)H^*(\lambda+\mu+\nu)\Phi(\lambda)\Phi^*(\lambda+\mu+\nu)\underline{T}^T(\nu)A_M^{-1}\underline{T}(\mu)\,d\lambda\,d\mu\,d\nu}{\frac{1}{2\pi}\int_{-\pi}^{\pi}|H(\lambda)|^2|\Phi(\lambda)|^2 d\lambda} \tag{3.94}$$

A comparison of (3.93) with (3.75) and (3.79) demonstrates that Theorem 3.6 is in agreement with the independently computed counterexample results. Moreover, we find that, in the general dependent case, the Mth order adaptive linear predictor's excess MSE is a linear function of the coefficient of kurtosis of the innovations process, thus confirming the conclusions drawn from the counterexamples.

Furthermore, examining (3.93) we see that the deviation of the excess error factor $\mu_D(M)$ in the general dependent case, from the value, M, it assumes in the Gaussian dependent case, is measured by the functional

$$C(H)\left[\frac{EW_n^4}{\sigma_W^4} - 3\right]. \tag{3.95}$$

Thus the deviation of the Mth order adaptive linear predictor's MSE, in the general dependent case, from the MSE incurred by the predictor in the Gaussian dependent case is dictated by this term. Hence the functional given in (3.95) deserves further consideration.

To this end, it can be shown that the magnitude of this functional is bounded, above and below, by the achievable bounds

$$0 \le |C(H)|\,\left|\frac{EW_n^4}{\sigma_W^4} - 3\right| \le M^{\frac{3}{2}}\frac{\lambda_{\max}}{\lambda_{\min}}\left|\frac{EW_n^4}{\sigma_W^4} - 3\right| \tag{3.96}$$

where $\lambda_{\min}$ and $\lambda_{\max}$ are the minimum and maximum eigenvalues of the processes' covariance matrix R_M, respectively.

The lower bound in (3.96) is achieved if the process to be predicted is either white (i.e., $H(\lambda) = 1$, $\lambda \in [-\pi, \pi]$) or Gaussian. Moreover, since the functional (3.95) is continuous in $H(\cdot)$, it follows that this lower bound is approached as the psd of the process becomes white.

Thus, as the process becomes linear mean square unpredictable, we find that the adaptive predictor's MSE approaches the value suggested by independent case based estimates such as the FPE statistic. Although this fact offers little insight into adaptive prediction applications of practical interest, it does nonetheless confirm our earlier observations regarding the apparent inaccuracy of the FPE statistic, on the basis of the computer generated results following Corollary 3.1.

Turning to the second factor

$$\left| \frac{EW_n^4}{\sigma_W^4} - 3 \right|$$

which determines the magnitude of the functional (3.95), we note that since the fourth central moment of a Gaussian random variable is precisely 3 times its variance, this factor may be viewed as a metric of the distance between the Gaussian and non-Gaussian cases. Indeed, Cramér (cf. [42], pp. 184-185) defines this factor as the *coefficient of excess* and observes that it is often taken as a measure of the deviation of a random variate's probability distribution from the normal distribution. Unfortunately, the magnitude of the coefficient of excess cannot be bounded without prior knowledge of the distribution in question.

In view of this, as well as the realization that the values of the minimal and maximal eigenvalues are determined by the (unknown) psd of the process to be predicted, we conclude that the deviation of the adaptive linear predictor's MSE from the value assumed in the dependent Gaussian case can be arbitrarily large in the non-Gaussian dependent case. This, when coupled with the fact that this deviation cannot be quantified without information on both the psd and the distribution of the innovations underlying the process to be predicted, constitutes the fundamental obstacle to *apriori* estimation of the adaptive linear predictor's MSE under the dependent case hypothesis.

3.2.3 Comparison with the Findings Kunitomo and Yamamoto

Our work would not be complete without a comparison of the present results with the significant findings of Kunitomo and Yamamoto [31]. However, as noted in section 3.1.4, the latter results, although impressive, are based on an AR process model and given in terms of infinite sums of the autocorrelation and autoregression coefficients of the process to be predicted. Consequently, a direct comparison of this theory with the analogous findings developed in Theorems 3.5 and 3.6 is prohibitively difficult. We thus restricted the desired comparison to the specific cases represented by counterexamples 1 and 2 of section 3.2.2.

Employing Kunitomo and Yamamoto's formulae, as found in Corollary 2 and Theorem 3 of their work, we obtained estimates of the adaptive linear predictor's MSE which were in exact agreement with the results given in (3.76) and (3.80). Hence it is likely that this agreement extends to our general theories as well.

Indeed, Kunitomo and Yamamoto appear to have been the first investigators to recognize the discrepancies between the independent and dependent case assumptions as well as the dependence of the adaptive linear predictor's MSE on the fourth cumulant[17] of the innovations process under the latter hypothesis. Since their work also revealed that this dependence is absent under the independent case assumption, Kunitomo and Yamamoto concluded, as we have, that independent case based estimates of the predictor's MSE are suspect. In this regard, our respective general theories can be considered to be in agreement with no uncertainty.

[17] Recall that the fourth cumulant is equal to the product of the coefficient of excess and the square of the variance.

3.3 Summary

In this chapter we have developed a general theory on the mean square error performance of the adaptive linear predictor under both the independent and dependent case hypotheses. As a consequence of these developments, it has been shown that independent case based estimates of the predictor's performance are unrepresentative of, and subject to significant error under, the conditions typically encountered in adaptive prediction environments and reflected by the dependent case hypothesis. The FPE statistic in particular was found not only to be inaccurate under the independent case assumption upon which it is predicated, but also to be intolerably misleading in non-Gaussian dependent case applications. Thus, independent case based performance estimation and order selection criteria, such as the Minimum FPE conjecture, were deemed unacceptable for these purposes and dismissed from further consideration.

The notoriety that the Minimum FPE conjecture has enjoyed is, perhaps, ironic in view of the conclusive evidence presented herein. Indeed, the conjecture is, in the dependent Gaussian case an exact restatement[18] of Davisson's order selection criterion or method, and in the independent case (under which its validity is claimed), or without the Gaussian assumption, it is simply false.

These findings clearly imply, and we therefore concluded, that any theory which is intended to accurately represent the conditions and performance expected in adaptive prediction applications must be derived under the dependent case model. The commitment to this more realistic model brought (as usual) a commensurate increase in the analytical difficulty of the problem - the reliable estimation of the adaptive linear predictor's optimal order and MSE performance. Nevertheless, it has been demonstrated that the fourth order moments which define the predictor's excess MSE, under such realistic conditions, are determined by the psd of the random sequence to be predicted as well as the coefficient of excess characterizing the probability distribution of the innovations process underlying this sequence. These functions are seldom known in adaptive prediction applications and their accurate estimation cannot be realized on the basis of the LMS residual or sample mean square prediction error alone.

Indeed, these difficulties were first encountered by Davisson [8] under the Gaussian dependent case hypothesis. Although the adoption of the Gaussian assumption eliminates the dependence of the adaptive predictor's excess MSE factor on the psd of the process in question, the predictor's MSE is nonetheless dependent upon this spectral density through the optimal predictor's MSE. Moreover, estimates of the optimal predictor's MSE based upon the LMS residual or sample mean square prediction error are subject to this dependence as well. Recognizing that these obstacles are not easily circumvented, Davisson had, in 1966, resigned to accept the consequential but tolerable error in his LMS residual based MSE estimation and order selection method for the adaptive prediction of Gaussian processes.

However, as demonstrated in section 3.2.2 and noted above, one must face the prospect of unbounded error when applying such techniques in the general dependent case so often encountered in adaptive prediction (i.e., when the Gaussian assumption cannot be justified). Clearly, a more robust performance estimation statistic and order selection criterion is needed.

[18] Albeit for the restricted special case of stationary Mth order Gauss-Markov processes.

Thus it seems that in this chapter we have managed to discredit all independent case based methods and cast suspicion upon known dependent case adaptive prediction performance estimates and order selection criteria. Unfortunately, this somewhat pessimistic assessment of the impediments to the accurate estimation of the predictor's performance and order, although advancing the theory, does not provide a solution to the subject problem in practical applications.

What then shall we do?

A potential (and quite possibly the only) solution is found in the problem's demonstration in Theorem 3.5. This theory proved that the mean square error incurred by the adaptive linear predictor in the dependent case is given, through terms of $O(\frac{M}{N})$, by

$$\sigma^2[M,N] = \sigma^2[M,\infty] + \frac{1}{N} EZ_n{}^2 \underline{S}_n^T R_M^{-1} \underline{S}_n \ . \tag{3.97}$$

Therefore the required solution would be effected if estimates of both the Mth order optimal predictor's MSE and the fourth order moments which constitute the excess MSE of the adaptive predictor of the same order, could be formulated which:

(1) possess desirable asymptotic as well as finite length realization attributes

and

(2) facilitate implementation in practical application environments (particularly in "real time" applications where on-line performance monitoring and order adaptation are required).

The discovery of this potential solution inspired the research presented in Chapter 4.

Chapter 4
An Extension of Davisson's Method

4.0 Introduction

The developments of Chapter 3 provide a definitive basis for the extension of Davisson's order selection criterion, under the dependent case hypothesis, to the adaptive linear prediction of processes which are not necessarily Gaussian. It is then logical that the formulation and analysis of such an extension are the subjects of the present chapter.

As in the preceding chapter, our analyses and theories are predicated on the premise that the process in question is stationary (at least locally) and strong mixing, with an associated mixing coefficient which meets certain sufficiency constraints. Thus, given our stated objective of extending the theory of adaptive linear prediction, it is appropriate that we begin this endeavor with an examination of the implications and limitations of these prerequisite hypothesis conditions. The strong mixing assumption in particular is analyzed to assess the constraints it imposes on the applicability of any theoretical extension based thereon.

Having characterized[1] the class of processes to which the required conditions of the hypothesis apply, we proceed, in section 4.2, to reexamine the adaptive linear predictor's MSE and the expected LMS fitting or sample mean square prediction error as given in Theorems 3.5 and 3.2, respectively. The asymptotic behavior of these expectations is scrutinized to both derive and intuitively justify the functional form of an estimator, of the adaptive linear predictor's MSE, that exhibits the potential of becoming unbiased with increasing sample size. The estimate thus formulated is shown, in section 4.3, to be asymptotically unbiased as postulated under the stationary, strong mixing hypothesis.

An extension of Davisson's adaptive linear predictor order selection criterion and MSE performance estimation theory follows as an immediate consequence of the demonstrated robustness of the extended estimator developed in sections 4.2 and 4.3. Indeed, we retain the logic and structure of Davisson's criterion, and simply propose the replacement of his original estimate (2.5), of the adaptive linear predictor's MSE, with the extended estimate as formulated herein.

This extension, or generalization of Davisson's theory does, however, exact a price in practical applications. The extended estimator is seen to be functionally more complex and hence computationally more intensive to evaluate than the original estimator which it is proposed to replace. Since computational complexity is often of paramount importance in on-line or real-time adaptive systems environments, we briefly examine the implementation requirements of the extended order selection criterion in such practical applications.

[1] To a limited extent.

4.1 Potential Limitations of Such an Extension

A fundamental question arising during most (if not all) theoretical modelling developments involves the extent to which the resulting theory applies in practice. The present research clearly provides no exception to this universal principle. Thus, in this section, we consider the degree to which the conditions common to the hypotheses of the prerequisite results found in Chapter 3 (cf. Theorems 3.2 and 3.5) as well as the developments contained in the sequel, limit the applicability of this theory.

Of primary interest are the potential limitations imposed by the underlying assumption that the process to be predicted is regular, stationary (strictly, at least over the duration of the learning period) and strong mixing, while manifesting a mixing coefficient meeting specific constraints. These issues are addressed in the following subsections.

4.1.1 The Necessity of Stationarity and Regularity

The assumption that the process to be predicted is strictly stationary, at least over the duration of the adaptation period, is the most benign and yet fundamental of the hypothesis conditions underlying the theory of this treatise. Indeed, without some degree of stationarity, the performance of any adaptive or learning system would not only be difficult to quantify, but also could, under a worst case assessment, be construed as meaningless. This observation necessarily follows when one recalls the rationale upon which adaptive systems are based. Conceptually, such systems are designed to learn the salient statistical features of their environments and, having learned (i.e., estimated) these features, to adapt their transfer functions according to some (typically pre-specified) control law. The implicit but nonetheless unavoidable assumption in this concept is that the statistical features in question remain effectively constant over the learning period as well as the interval in which the system's adapted response is applied.

In the absence of stationarity over the learning period, the accuracy of the system's feature estimates (which are usually parametric) and the applicability of the induced (adapted) transfer function are suspect at best. (And here we have not considered the potential and additional degradation arising in non-stationary environments where the learning period and the interval over which the adapted response is applied do not coincide. These intervals may in fact be separated by a considerable delay due to operational constraints.) The necessity of stationarity in the context of an adaptive system is thus quite clear.

Fortunately, the effective period of stationarity is ascertainable with minimal knowledge of the application environment. A rudimentary understanding of the underlying physical mechanisms which give rise to the observations and insight into the sampling characteristics of the instrumentation system in question are often sufficient to conservatively determine the maximal period of effective stationarity.

In view of the preceding arguments we conclude that the assumption of stationarity does not unduly limit the applicability of the theory herein.

We now consider the necessity of and potential limitations resulting from the assumption that the process to be predicted is regular. We begin by recalling (cf. section 1.1) that a stochastic process is said to be *regular*[2] if $\sigma^2[M,\infty] > 0$ for all $1 \le M \le \infty$; that is, if the process cannot be predicted with complete certainty, in a mean square error sense, by a linear predictor structure. As such, this property has, as noted in Chapter 1, been assumed to hold throughout this treatise. This fact remains even though regularity is not explicitly stated as a prerequisite condition in the hypotheses of results requiring the strong mixing condition (cf. Theorems 3.2, 3.5, and 4.1) - for as shown in the sequel, regularity necessarily follows from the strong mixing property. Clearly then, the implications of regularity are of interest here.

Perhaps the most commonly recognized consequence of regularity (cf. Appendix 1A) is the fact that it guarantees the positive definiteness of the processes' covariance matrix R_M for all $M \geq 1$. This in turn ensures the existence and uniqueness of the Mth order optimal predictor, thereby facilitating the analyses of Chapter 3, and sections 4.2 and 4.3. Moreover, several authors [40], [49], and [50] have shown that a stationary discrete time parameter process is regular if, and only if, it is representable as either an "infinite" MA or "infinite" AR process:

$$S_n = \sum_{k=0}^{\infty} a_k \epsilon_{n-k} \quad \text{with} \quad \sum_{k=0}^{\infty} |a_k|^2 < \infty \tag{4.0}$$

or

$$S_n = \sum_{k=1}^{\infty} b_k S_{n-k} + \epsilon_n \quad \text{with} \quad \sum_{k=1}^{\infty} |b_k|^2 < \infty$$

where $\{S_n\}$ is the regular process in question and $\{\epsilon_n\}$ represents the usual uncorrelated or orthogonal innovations sequence.

This result clearly implies that all stationary MA and AR processes are regular and thus lends intuitive credibility to the use of a linear autoregressive structure for predicting such processes.

The most insightful necessary and sufficient condition for regularity however, is due to Doob [40] and discussed by Rozanov [49] as well as Ibragimov and Linnik [50]. Doob demonstrates that a process with a power spectral density[3] (psd) $P_{SS}(\lambda)$ is regular if, and only if,

$$\int_{-\pi}^{\pi} \log P_{SS}(\lambda) d\lambda > -\infty \tag{4.1}$$

and is *deterministic* or *singular* otherwise.

Thus a regular process must admit a psd that is almost everywhere nonzero, and which, does not vanish too rapidly on any set of positive measure within the interval $[-\pi, \pi]$. But given the imperfect filtering and finite sampling rates inherent in realizable digital processing systems, it seems intuitively obvious that (4.1) is readily satisfied in most if not all applications of practical interest. This, when coupled with the regularity of the comprehensive class containing all stationary linear processes, leads us to conclude that the property of regularity can be expected to be encountered in essentially all adaptive prediction environments.

Unfortunately, such a conclusion is not quite so obvious when considering the potential restrictiveness of the strong mixing property, as seen in the following section.

---

[2] Rozanov [49], and Ibragimov and Linnik [50] refer to such processes as *linearly regular* while defining mixing processes as *regular*, and processes possessing the property of strong mixing as *completely regular*.

[3] The existence of the psd is shown by Ibragimov and Linnik, [50] pp. 303, to constitute a necessary condition for regularity. (4.1) is often cited as the necessary and sufficient Paley-Wiener condition for (4.0).

4.1.2 Implications of the Strong Mixing Hypothesis

The theory of *mixing* or *weakly dependent* stochastic processes has received considerable attention [43-51] since the advent of the first mixing condition, as proposed by Rosenblatt [43] 35 years ago. However, in spite of this apparent interest, few processes "... have been characterized in terms of ergodic or mixing properties " as Rosenblatt has more recently noted (cf. [51] pp. 112). This unfortunate fact can undoubtedly be attributed to the difficulty of the subject.

In view of these apparent difficulties, we shall confine the present analysis to the interpretation of the body of knowledge available in the literature - for an independent study of this field is clearly beyond the scope of this treatise. With this caveat, we proceed to consider the implications of the strong mixing hypothesis.

Although numerous mixing conditions have been defined, three forms have been widely studied: *symmetric mixing* due to Philipp[4], *uniform* or ϕ *mixing* as proposed by Ibragimov and studied extensively by Billingsley [46], and *strong mixing* as attributed to Rosenblatt [43], and adopted by the present work. As noted in section 3.1.2, a stationary discrete time parameter stochastic process $\{S_n\}_{n=-\infty}^{+\infty}$ is said to satisfy a *strong mixing* condition if there exists a monotonically decreasing sequence $\alpha_k \downarrow 0$ as $k \to \infty$ such that

$$|P(A \cap B) - P(A)P(B)| \leq \alpha_k \tag{4.2}$$

for all $A \in M_{-\infty}^{n}$, $B \in M_{n+k}^{+\infty}$ and all $0 \leq k < \infty$ where $M_{n_1}^{n_2}$ denotes the σ algebra of events generated by the random variables $\{S_{n_1}, \cdots, S_{n_2}\}$. The symmetric and uniform mixing conditions are defined[5] in a similar but more restrictive manner.

The latter two mixing conditions are mentioned only to note that the adjective *strong*, describing the first mixing condition, is somewhat of a misnomer. Indeed, it is easily shown [47] that the property of symmetric mixing necessarily implies satisfaction of the uniform mixing condition, which in turn, implies the property of strong mixing. Thus, in this sense, the theory of the present treatise is based upon the least restrictive of these commonly invoked mixing conditions. We forego further discussion of the former two conditions to concentrate our attention on the strong mixing condition, as this property is germane to our work.

Ibragimov and Linnik [50], *et al*, have shown that any strong mixing process is necessarily mixing, and, further that any second order mixing process must be regular. Thus, satisfaction of the strong mixing condition implies regularity and hence (4.1) constitutes a condition necessary for the strong mixing property. An alternate but analogous form of this result follows as a consequence of Lemma 3.1 (Davydov's lemma). Applying this lemma we find that the magnitude of a strong mixing processes' autocorrelation function, R_k, must satisfy

$$|R_k| \leq 8\alpha_k^{\frac{\delta}{2+\delta}} \{E\,|S_0|^{2+\delta}\}^{\frac{2}{2+\delta}} \quad \text{for } all\ k \geq 0 \text{ and } some\ \delta > 0\,. \tag{4.3}$$

Since, from (4.2), $\alpha_k \downarrow 0$ as $k \to \infty$, then (4.3) dictates the minimal rate at which a strong mixing processes' autocorrelation function must decay with increasing intersample delay.

---

[4] W. Philipp, "The Central Limit Problem for a Mixing Sequence of Random Variables," Z. Wahrscheinlichkeitstheorie und verw. Geb., vol. 12, pp. 155-171, 1969.

[5] The details are omitted here since these conditions are not relevant to the present work.

This, when taken with the reciprocal spreading theorem of Fourier transform theory, is equivalent and lends intuitive justification to (4.1) as a necessary condition for strong mixing.

As previously noted, (4.1) represents a relatively mild constraint in that its satisfaction is anticipated in typical adaptive prediction applications. Thus the conditions given in (4.1) or (4.3) do not imply that the strong mixing assumption is unduly restrictive in this context. On the other hand, these conditions are only necessary.

Unfortunately, conditions which are both necessary and sufficient to ensure that a stationary process is strong mixing, and easily verified in practice, are apparently not known. Nevertheless, a substantial collection of specific classes of stationary processes has been shown[6] to satisfy the strong mixing condition. This collection includes all:

(i) i.i.d. processes

(ii) m-dependent processes (including MA processes derived from i.i.d. innovations)

(iii) processes defined as measurable functions of a finite number of the variates of a stationary strong mixing process; i.e.,

$$Y_n = f(X_{n+n_1}, X_{n+n_2}, \cdots, X_{n+n_m})$$

where $m < \infty$ and $\{n_1, \ldots, n_m\}$ are fixed, and $\{X_n\}_{n=-\infty}^{+\infty}$ is stationary strong mixing (Billingsley [46], Ibragimov and Linnik [50]).

(iv) Markov chains with transition probabilities meeting certain constraints (Ibragimov and Linnik [50], pp. 365-369 and 420-424).

(v) Gaussian processes admitting continuous psds $P_{SS}(\lambda)$ such that $P_{SS}(\lambda) \geq \epsilon > 0$ for all $\lambda \in [-\pi, \pi]$ (Kolmogorov and Rozanov [48]).

We examine the conditions which have been shown to be necessary and sufficient for stationary Gaussian processes to possess the strong mixing property, since the restrictiveness of such an assumption, relative to Davisson's regular, ergodic, Gaussian hypothesis [7-8], is of particular interest. To facilitate the derivation of these necessary and sufficient conditions, Kolmogorov and Rozanov [48] defined the *maximal correlation coefficient* of a second order stationary discrete time parameter process as

$$\rho_k = \sup_{X,Y} \frac{E(X - EX)(Y - EY)}{\{E(X - EX)^2 E(Y - EY)^2\}^{1/2}}$$

where X,Y are any measurable functions (random variables) with finite second moments defined on $M_{-\infty}^{n}$ and $M_{n+k}^{+\infty}$, respectively. With this definition, these noted probability theorists formulated what is now known as the Kolmogorov-Rozanov Theorem:

Given any stationary second order Gaussian process, then

$$\alpha_k \leq \rho_k \leq 2\pi\alpha_k \quad , \quad 0 \leq k < \infty \tag{4.4}$$

and hence a stationary second order Gaussian process is strong mixing if, and only if,

$$\lim_{k \to \infty} \rho_k = 0 . \tag{4.5}$$

[6] That the first two classes are strong mixing follows trivially from the fundamental definition, (4.2), of this property.

Digressing momentarily, it is informative to note that the inequality on the left in (4.4), $\alpha_k \leq \rho_k$, holds for any stationary, second order, regular process. Hence all processes satisfying a maximal correlation condition (i.e., processes which are asymptotically uncorrelated as measured by the satisfaction of (4.5)) are necessarily strong mixing as well. Thus we find that the strong mixing condition is implied by yet another (more restrictive) form of asymptotic independence.

Kolmogorov and Rozanov further demonstrated that the continuous, non-vanishing psd condition (given under (v) above) is sufficient to guarantee (4.5) and thus that the Gaussian process in question is strong mixing[7]. Exploiting this condition, Ibragimov and Linnik [50] defined an "extensive" class of Gaussian processes which are strong mixing, while Rozanov [49] notes that when restricted to stationary Gaussian processes, the properties of regularity and mixing are equivalent.

Continuity of the psd on $[-\pi, \pi]$, as dictated by the Kolmogorov-Rozanov sufficiency condition (v), implies that the psd is bounded and hence square integrable on this interval. Moreover, (v) requires that the psd is strictly positive over this domain which clearly implies the satisfaction of (4.1) and hence regularity. Thus, satisfaction of Davisson's hypothesis requiring both regularity and the ergodic constraint $P_{SS}(\lambda) \in L_2[-\pi, \pi]$ necessarily follows from the satisfaction of the Kolmogorov-Rozanov sufficiency condition. This however does not suggest that a strong mixing Gaussian process meets Davisson's assumptions - for the Kolmogorov-Rozanov sufficiency condition has not been shown to be necessary. Conversely, square integrability of the psd has not been shown to be either necessary or sufficient to ensure the strong mixing property in Gaussians. The relative restrictiveness of the strong mixing assumption and Davisson's hypothesis is therefore difficult to assess precisely.

Nonetheless, it is, as noted above, possible to construct a large class of Gaussian processes which are strong mixing and satisfy Davisson's hypothesis. In this sense it is not unreasonable to argue that Davisson's hypothesis and the strong mixing condition impose comparable limitations on the applicability of the theory following from their adoption. Furthermore, our claim of extending Davisson's theory derives from the fact that the strong mixing hypothesis, and hence the main results of this treatise, (i.e., Theorems 3.2, 3.5, and 4.1), are applicable in environments where the Gaussian assumption is not.

The astute reader will have, at this point, noticed that we have not addressed the limitations potentially imposed by the remaining condition prerequisite to Theorems 3.2, 3.5, and 4.1. These results require, in addition to the strong mixing property, that for some $\delta > 0$,

$$\sum_{k=1}^{\infty} \alpha_k^{\frac{\delta}{2+\delta}} < \infty \quad \text{and} \quad E\,|S_0|^{4(2+\delta)} < \infty . \tag{4.6}$$

where α_k represents the strong mixing coefficient of the process in question. This condition obviously imposes further constraints on the rate at which α_k must vanish as $k \to \infty$ (i.e., the rate at which samples from the process become independent as the delay between the samples increases without bound) as well as the higher order absolute moments of the process.

[7] This result, when compared with (4.1), perhaps explains why some authors refer to the property of *strong mixing* as *completely regular*.

Indeed, employing the well known Comparison test for determining the convergence (or divergence) of an infinite series of non-negative terms, with the p- series $\sum_{k=1}^{\infty} k^{-p}$ (for comparison), we find that (4.6) is satisfied if, and only if,

$$\alpha_k \leq k^{-(1+\frac{2}{\delta})p} \quad \text{and} \quad E\,|S_0|^{4(2+\delta)} < \infty \quad \text{for } some\ p > 1 \text{ and } \delta > 0\,. \tag{4.7}$$

Examining (4.7) reveals that (4.6) is potentially satisfied by a range of values for δ. Selecting the minimal value of δ within this range minimizes the order of the absolute moments which must be finite. If, for example, a value of $\delta << 1$ is sufficient to meet the convergence criterion on α_k in (4.7) (thus indicating that the dependence between samples from the process diminishes rapidly as the number of sampling periods, k, separating them increases) then (4.7) only requires the eighth order moments of the process to be finite. On the other hand, if a large value of δ (say $\delta >> 1$) is required to satisfy the convergence criterion (thereby indicating that the process becomes asymptotically independent at a slower rate), then (4.7) implies that proportionally higher order moments must be finite to ensure the satisfaction of (4.6). In either case, the implicit constraint on the psd is intuitively consistent with (4.1) and (4.3). Moreover, the finite dynamic range inherent in realizable digital systems guarantees the finiteness of the absolute moments as required by (4.6) and (4.7).

Hence, we conclude that (4.6) should be satisfied by essentially all processes that are to be encountered in adaptive prediction applications.

Although we have demonstrated that the strong mixing condition (4.2) and the constraints (4.6) associated with it are likely to be met in most adaptive prediction environments, we regrettably have not provided a sufficiency test for verifying the same. This shortcoming has been the source of considerable dismay, since the strong mixing hypothesis and the associated constraints (4.6) were adopted[8] with the objective of basing our main results (Theorems 3.2, 3.5, and 4.1) on conditions which would be readily verifiable in practice. The only consolation in this regard comes with the realization that renowned probability theorists such as Kolmogorov, Rosenblatt, Rozanov, Ibragimov, and Linnik have apparently been unable to do the same. Indeed, in a section entitled "Some Unsolved Problems" , [50], pp. 393, Ibragimov and Linnik observe that "A vast field of research is presented by the problem of characterizing stationary processes satisfying one or other of the conditions of weak dependence. What conditions, for example, are laid upon the moments of a stationary non-Gaussian process by the strong mixing condition? " Thus it seems that our expectations, upon adopting the strong mixing condition, exceeded the limited scope of the present research.

One might ask, at this juncture, why we would choose to adopt a hypothesis entailing conditions that, in retrospect, are difficult to verify in practice. This in turn suggests that one question the very necessity of the strong mixing condition as a prerequisite to the results of Theorems 3.2, 3.5, and 4.1. For, although this theory clearly demonstrates the sufficiency of the strong mixing hypothesis in guaranteeing the subject results, the question of necessity nonetheless remains.

On this issue we can only note that the property of ergodicity is necessary in any adaptive or learning system environment. In discussing the necessity of ergodicity in such practical situations, Gray and Davisson, [45] pp. 171, observe that "The crucial point for applications is the knowledge that the sample average will ... converge to *something* (and hence the observer can in principle learn it) ...".

[8] This research initiative obtained earlier results under a more restrictive and arcane mixing condition

In this context, Davisson and others, [44] pp. 132, have shown that a stationary strong mixing process is ergodic and satisfies one or more ergodic theorems (cf. Lemma 3.2). Furthermore, we can intuitively argue that in theoretical studies such as the one at hand, where the rate of convergence of the sample averages is as critical as the fact that convergence occurs, a mixing condition which ensures a specific convergence rate and permits this rate to be quantified, is essential. Thus, although we have not answered the question of necessity, we assert that the strong mixing property is the least restrictive mixing condition which ensures the necessary property of ergodicity, and facilitates the determination of the convergence rate of sample statistics. In this respect at least, the strong mixing process is thought to be necessary.

4.2 On Formulating an Extended Estimator of the Adaptive Linear Predictor's MSE

In this section we develop an estimator for the mean square error incurred by the adaptive linear predictor, under the dependent case hypothesis, in environments where the Gaussian assumption is not necessarily applicable. This estimator is designed, by employing the results of Theorems 3.2 and 3.5, to become unbiased asymptotically, as the duration, N, of the predictor's learning period becomes infinite. Since Theorems 3.2 and 3.5 rely only upon the stationary strong mixing hypothesis, the estimator formulated below provides a basis for an extension of Davisson's order selection criterion and performance estimation theory under these conditions.

The development of such an extended estimator begins by noting that the mean square prediction error $\sigma^2[M,N]$ of the Mth order adaptive linear predictor is, in the dependent case, found from Theorem 3.5 to be

$$\sigma^2[M,N] = \sigma^2[M,\infty] + \frac{1}{N}EZ_n^2\underline{S}_n^T R_M^{-1}\underline{S}_n + o(\frac{M}{N}) . \tag{4.8}$$

However, as demonstrated in section 3.2.2, the magnitudes of the optimal predictor's MSE $\sigma^2[M,\infty]$ and the adaptive predictor's excess error $\frac{1}{N}EZ_n^2\underline{S}_n^T R_M^{-1}\underline{S}_n$ cannot be bounded *a priori* beyond the trivial observation that these performance measures are finite. Thus, estimates of these quantities must be formulated as the only reliable means of estimating $\sigma^2[M,N]$ accurately.

To this end we note that the mean LMS fitting or residual error is, under the stationary strong mixing hypothesis, given by Theorem 3.2 as

$$EF^2(M,N) = \sigma^2[M,\infty] - \frac{1}{N^2}\sum_{m=-N}^{N}(N-|m|)EZ_0Z_m\underline{S}_0^T R_M^{-1}\underline{S}_m + o(\frac{M}{N}) . \tag{4.9}$$

Expanding this result as

$$EF^2(M,N) = \sigma^2[M,\infty] - \frac{1}{N}\sum_{m=-N}^{N}EZ_0Z_m\underline{S}_0^T R_M^{-1}\underline{S}_m + \frac{2}{N^2}\sum_{m=1}^{N}mEZ_0Z_m\underline{S}_0^T R_M^{-1}\underline{S}_m + o(\frac{M}{N})$$

and noting that the third term on the right is shown, in Appendix 3D, to be $o(\frac{M}{N})$ under the stationary strong mixing hypothesis, we find that

$$EF^2(M,N) = \sigma^2[M,\infty] + \frac{1}{N}EZ_0^2\underline{S}_0^T R_M^{-1}\underline{S}_0 - \frac{2}{N}\sum_{m=0}^{N}EZ_0Z_m\underline{S}_0^T R_M^{-1}\underline{S}_m + o(\frac{M}{N}) . \tag{4.10}$$

Comparing (4.8) and (4.10) suggests that an asymptotically unbiased estimate of the adaptive linear predictor's MSE can be realized by

$$\hat{\sigma}^2[M,N] = F^2(M,N) + \frac{2}{N}Q(M,N) \tag{4.11}$$

if $Q(M,N)$ is in fact formulated as an asymptotically unbiased estimate of the unknown moments

$$\sum_{m=0}^{N} EZ_0 Z_m \underline{S}_0^T R_M^{-1} \underline{S}_m \ .$$

The logical if not intuitively obvious choice for an asymptotically unbiased estimator of these moments is found in their corresponding sample mean:

$$Q(M,N) = \frac{1}{N}\sum_{k=1}^{N}\sum_{l=0}^{N}(S_{n-k} - \underline{\beta}_{M,N}^T \underline{S}_{n-k})(S_{n-k-l} - \underline{\beta}_{M,N}^T \underline{S}_{n-k-l})\underline{S}_{n-k}^T r_M^{-1} \underline{S}_{n-k-l} \tag{4.12}$$

where of course, $\underline{\beta}_{M,N}$ represents the Mth order adaptive predictor's coefficients.

Thus we claim that $\hat{\sigma}^2[M,N]$, as defined by (4.11) and (4.12), constitutes an asymptotically unbiased estimator for the Mth order adaptive linear predictor's MSE. A proof of this assertion is given in the following section.

4.3 An Asymptotically Unbiased Estimator of the Adaptive Linear Predictor's MSE

With the developments of the preceding section, we can now establish one of the main results of this chapter - an extended, asymptotically unbiased estimator of the adaptive linear predictor's mean square error. This result is given in the theorem which follows.

Theorem 4.1

Given a zero mean, stationary, strong mixing process, $\{S_n\}_{n=-\infty}^{+\infty}$, with mixing coefficient α_k such that

$$\sum_{k=1}^{\infty} \alpha_k^{\delta^*} < \infty \quad \text{and} \quad E\,|S_0|^{4(2+\delta)} < \infty \quad \text{for } some \ \delta > 0$$

where $\delta^* = \dfrac{\delta}{2+\delta}$, then an asymptotically unbiased estimator for the MSE, $\sigma^2[M,N]$, incurred by the Mth order adaptive linear predictor under the dependent case hypothesis is given by

$$\hat{\sigma}^2[M,N] = F^2(M,N) + \frac{2}{N}Q(M,N) \tag{4.13}$$

where

$$Q(M,N) = \frac{1}{N}\sum_{k=1}^{N}\sum_{l=0}^{N}(S_{n-k} - \underline{\beta}_{M,N}^T \underline{S}_{n-k})(S_{n-k-l} - \underline{\beta}_{M,N}^T \underline{S}_{n-k-l})\underline{S}_{n-k}^T r_M^{-1} \underline{S}_{n-k-l} \tag{4.14}$$

with r_M, $\underline{\beta}_{M,N}$, and $F^2(M,N)$ as defined in (1.4) and (1.5) respectively. Furthermore, under the stated conditions,

$$E\hat{\sigma}^2[M,N] = \sigma^2[M,N] + o\left(\frac{M}{N}\right) \tag{4.15}$$

where $\lim_{\frac{N}{M}\to\infty} \frac{N}{M} o\left(\frac{M}{N}\right) = 0$.

Theorem 4.1 provides the essential prerequisite for an extension of Davisson's adaptive linear predictor order selection criterion and performance estimation theory to applications where non-Gaussian statistics may be encountered - an estimator for $\sigma^2[M,N]$ that is asymptotically unbiased in such environments. However, a review of Davisson's Method (cf. section 2.1.2) will reveal that this criterion also employs an asymptotically unbiased estimate of the Mth order optimal predictor's MSE (to support the determination of the adaptive predictor's optimal learning period). Fortunately, such an estimator follows, under the hypothesis, and as an immediate consequence of, Theorem 4.1. We therefore provide the following corollary which will prove useful in extending Davisson's Method.

Corollary 4.1

Under the conditions of Theorem 4.1, an asymptotically unbiased estimate of the Mth order optimal predictor's MSE is given by

$$\hat{\sigma}^2[M,\infty] = F^2(M,N) + \frac{2}{N}Q(M,N) - \frac{1}{N}Q_0(M,N) \tag{4.48}$$

where

$$Q_0(M,N) = \frac{1}{N}\sum_{k=1}^{N}(S_{n-k} - \underline{\beta}_{M,N}^T \underline{S}_{n-k})^2 \underline{S}_{n-k}^T r_M^{-1} \underline{S}_{n-k} \,. \tag{4.49}$$

With the benefit of Theorem 4.1 and Corollary 4.1, we are in a position to extend Davisson's order selection criterion and performance estimation theory, for adaptive linear predictors, beyond the constraints inherent in the assumption of Gaussian statistics. This extension is thus the subject of the following section.

4.4 Davisson's Method Extended

It is widely recognized that the performance of an adaptive linear predictor or AR model is critically dependent upon model order. Adopting an adaptive predictor with an order that is either too small (i.e., "underfitting") or large ("overfitting") results, as shown in Chapter 3, in excessive mean square prediction error. Moreover, in those applications where the adaptive linear predictor's coefficients are computed to support autoregressive spectral estimation, model underfitting results in underresolving, while model overfitting induces spurious detail (i.e., peaks) in, the resulting spectral estimate. Thus selecting the adaptive linear predictor or AR model order that constitutes an optimal balance between faithfully representing the (unknown) autocorrelation characteristics of the process in question, and the detrimental effect of basing an over-parameterized model on a finite, N, length observation sequence (learning period), is crucial.

The importance of proper model order selection is, in practical applications, perhaps exceeded only by the need to accurately estimate the performance of the resulting adaptive predictor. In practice, potential applications are often accompanied by a specification of the minimum acceptable performance level(s) that must be achieved by the adaptive predictor under consideration. Thus, the decision to implement an adaptive structure hinges, in many cases, upon the anticipated performance of such a system. A reliable means of estimating the expected performance of an adaptive linear predictor therefore becomes a prerequisite to its implementation.

The need for a reliable estimate of the predictor's performance also exists in applications where an adaptive structure is dictated by operational considerations (e.g., nonstationarity) and is therefore mandated without preconditions on its performance. For in these, as in all, applications, the task of designing the required predictor, including the selection of its order, remains. But all well known order selection criteria, such as Davisson's Method, are based on estimates (albeit of varying degrees of accuracy) of the candidate predictor's performance. The common strategy for adaptive linear predictor order selection is indisputably logical - select that order which results in the best expected performance, as measured by the estimator upon which the criterion is based. The applicability and accuracy of such order selection criteria are thus ultimately determined by the limitations and reliability of the underlying adaptive linear predictor performance estimate.

Davisson's Method represents the most widely recognized[9] adaptive linear predictor order selection criterion and performance estimation theory currently in use. Nevertheless, Davisson's Method is not, as demonstrated in Chapters 2 and 3, immune to the limitations of the adaptive linear predictor performance estimate upon which it is based. Indeed, Davisson's estimate of the adaptive predictor's MSE (2.5) is predicated on the assumption that the process to be predicted is Gaussian. Unfortunately, deviation from this hypothesis can, in practice, lead to significant degradation in the accuracy of this estimate (cf. section 3.2.2). Moreover, as Davisson originally noted in 1966, and is shown in section 3.2.2, this estimator is biased[10] even when applied in Gaussian prediction environments.

In view of the preceding arguments, the need for an accurate and robust estimate of the adaptive linear predictor's MSE performance is quite clear. The recognition of this need motivated the research which culminated in the formulation of the estimate (4.13) of the adaptive predictor's MSE. This statistic constitutes an asymptotically unbiased estimate of the mean square error produced by an adaptive linear predictor, when applied to any strong mixing process admitting a mixing coefficient meeting the constraints specified in Theorem 4.1. Furthermore, as discussed in section 4.1.2, the class of processes satisfying the conditions of Theorem 4.1 includes not only a substantial collection of Gaussians but also non-Gaussian processes as well. The estimates given in Theorem 4.1 and Corollary 4.1 thus facilitate an extension of Davisson's Method beyond the limitations imposed by the Gaussian hypothesis.

To extend Davisson's Method we retain the logical procedure of the original order selection criterion by simply replacing Davisson's estimates of the Mth order adaptive and optimal predictors' mean square errors with those derived in Theorem 4.1 and Corollary 4.1, respectively. The criterion that results, for selecting the order and estimating the MSE performance of an adaptive linear predictor under the dependent case hypothesis, can therefore be stated as follows:

<u>Davisson's Method Extended</u>

(i) The adaptive linear predictor's optimum memory, or order, M_0, is given by that value of M which minimizes $\hat{\sigma}^2[M,N]$ over $1 \leq M \leq M_{\max}$ where

$$\hat{\sigma}^2[M,N] = F^2(M,N) + \frac{2}{N} Q(M,N) \tag{4.52}$$

with the LMS fitting error, or residual, $F^2(M,N)$ as defined in (1.5),

$$Q(M,N) = \frac{1}{N} \sum_{k=1}^{N} \sum_{l=0}^{N} (S_{n-k} - \underline{\beta}_{M,N}^T \underline{S}_{n-k})(S_{n-k-l} - \underline{\beta}_{M,N}^T \underline{S}_{n-k-l}) \underline{S}_{n-k}^T r_M^{-1} \underline{S}_{n-k-l} \quad ,$$

[9] The fact that many authors refer to this criterion as the Minimum FPE Procedure and attribute it to Akaike, notwithstanding.

[10] This bias results from the use of a legitimate but biased approximation of the expected LMS fitting error.

and M_{max} and N are selected such that $M_{max} << N < N_{max}$ (N_{max} is both determined and upper bounded by the effective period of stationarity for the process to be predicted).

(ii) The optimum learning or adaptation period, N_0, is subsequently given by the minimum value of N satisfying

$$\frac{\hat{\sigma}^2[M_0,N]}{\hat{\sigma}^2[M_{max},\infty]} \leq 1+\epsilon \tag{4.53}$$

where, from (4.48) of Corollary 4.1

$$\hat{\sigma}^2[M_{max},\infty] = \hat{\sigma}^2[M_{max},N_{max}] - \frac{1}{N_{max}} Q_0(M_{max},N_{max}) \tag{4.54}$$

and ϵ represents the acceptable percentage by which the M_0th order adaptive predictor's MSE is allowed to exceed that of the optimal predictor of maximal order. (Selection of the optimum learning period, N_0, under this criterion minimizes the degrading effects of potential process nonstationarities).

(iii) An asymptotically unbiased estimate of the mean square prediction error produced by the M_0th order predictor specified in (i) and (ii), is given by $\hat{\sigma}^2[M_0,N_0]$ (i.e., (4.52) evaluated at M_0, N_0).

The use of this criterion is, of course, subject to the fundamental constraints and considerations that inherently characterize adaptive linear prediction applications. The maximal period of effective stationarity, N_{max}, must be conservatively estimated (cf. section 4.1.1) for the process to be predicted, and the input (observation) sequence is subsequently partitioned into contiguous learning periods (frames) of length $N \leq N_{max}$. The adaptive linear predictor's coefficients are typically estimated (updated) on the basis of, and applied to, the data within each frame, independent of the observations within preceding frames.

Since it is well known that the adaptive predictor's excess MSE is proportional to the ratio of its order to the length of its learning period (i.e., $\frac{M}{N}$, see for example, Chapter 3), the maximal admissible predictor order, M_{max}, must be constrained to be much less than the length of the learning period to minimize the effect of this source of performance degradation. As a consequence of restricting $M_{max} << N$, the magnitude of the $o(\frac{M}{N})$ terms in the asymptotic results of Theorems 3.2, 3.5, 4.1, as well as Corollary 4.1, become insignificant relative to the order of the terms that are retained. The bias induced by neglecting these terms in our finite length learning period approximations should thus be negligible.

Inasmuch as it is possible that the "conservative" estimate of the effective period of stationarity, N_{max}, may in fact exceed the period over which the input statistics remain effectively constant, selecting the optimum learning period, N_0, as the adaptation period of minimum length that achieves the desired performance measure given in (4.53) and (4.54), further mitigates the effect of this potential source of performance loss. However, is should be noted that the length of the learning period and the computational intensity induced by the resulting adaptation algorithm share an inverse relationship - minimizing the length of the learning period maximizes the frequency with which the adaptive predictor's coefficients must be updated, and hence the computational resources required to do so. Thus a compromise, between the adaptive predictor's immunity to potential process nonstationarities and the computational throughput required to implement the predictor, may in fact be necessary.

The computational requirements of the adaptive predictor are intensified in those real-time environments that demand on-line estimation of model order (e.g., prediction of multiple, spectrally diverse sources). In such applications the predictor's order must be periodically updated (adapted), thus necessitating an on-line implementation of the order selection criterion. The order update period is dictated by the duration of process order (source) stationarity, but in any case is restricted to be an integral multiple of the predictor's coefficient learning period. Thus it is entirely possible that the required order update period may be short (i.e., a small multiple of the coefficient learning period) in environments with either highly dynamic or multiple, intermittent sources. The computational resources necessary to support on-line model order adaptation in these environments can, depending on the order selection criterion in question, be substantial.

Unfortunately, the computational requirements implied by a direct implementation of Davisson's Method Extended clearly exceed those of Davisson's original method, and may in fact be greater than the computational burden of order selection methods such as the PLS criterion [35]. The latter criterion minimizes the inherent computational redundancy of the LMS algorithm by employing computationally efficient predictive lattice filters. Thus we believe that a significant reduction in the implied computational requirements of Davisson's Method Extended might be realized by a judicious application of similar implementation techniques.

5.3 Areas Requiring Further Research

It is somewhat humbling to acknowledge that, in spite of our considerable and continuing efforts, several issues remain to be resolved. These include:

1. A more definitive characterization of the class of processes satisfying the strong mixing conditions of Theorems 3.2, 3.5, and 4.1.
2. An analysis of the minimal additive order penalty term necessary and sufficient to ensure the strong consistency of Davisson's Method Extended as an estimator of process order. On this issue, we must reiterate (cf. section 1.1) that consistency only becomes meaningful if the process to be predicted in fact admits a (finite) AR representation (and many processes of course do not). Thus forcing consistency through the addition of an order penalty term may degrade the MSE performance of the adaptive predictor by unnecessarily penalizing higher orders which, relative to the MMSE criterion, would be chosen otherwise. We nevertheless believe that the issue of consistency should be addressed in view of the emphasis placed on this asymptotic property by many authors.
3. An evaluation of the relative accuracy of order estimation of Davisson's Method Extended, and the AIC, BIC/MDL, CAT, PLS, HQ, and Davisson's (original) criterion, when the process in question is not necessarily Gaussian.
4. A reduction of the computational requirements (implied by a direct implementation) of Davisson's Method Extended by exploiting the structure and inherent redundancy of the method's LMS calculations.

As this research initiative was undertaken and executed with considerable interest, we plan to continue to pursue our investigation of these issues. In the immortal words of David Hilbert, " We hear within us the perpetual call; There is the problem. Seek its solution.". And so we shall.

Acknowledgements

The authors wish to express their gratitude to Dr. Jorma Rissanen, of the IBM Almaden Research Center, for bringing the publications of Akaike to their attention, as well as his encouragement in our initial investigation of this theory. We are also indebted to Professor Ching-Zong Wei, of the University of Maryland, for providing invaluable references in the literature, and Professor Laszlo Gyorfi, of the Technical University of Budapest, for his careful review of our manuscript.

References

[1] L. D. Davisson, "Theory of Adaptive Data Compression," Ph.D. Dissertation, University of California at Los Angeles, 1964.

[2] J. Makhoul, "Linear Prediction: A Tutorial Review," *Proc. IEEE,* vol. 63, no. 4, pp. 561-580, April 1975.

[3] B. Widrow *et al.*, "Adaptive Noise Cancelling: Principles and Applications," *Proc. IEEE,* vol. 63, no. 12, pp. 1692-1716, Dec. 1975.

[4] F. W. Symons, "Narrow-Band Interference Rejection Using The Complex Linear Prediction Filter," *IEEE Trans. Acoust., Speech, Signal Processing,* vol. ASSP-26, pp. 94-98, Feb. 1978.

[5] S. M. Kay and S. L. Marple, Jr., "Spectrum Analysis- A Modern Perspective," *Proc. IEEE,* vol. 69, no. 11, pp.1380-1419, Nov. 1981.

[6] L. B. Milstein, "Recent Developments In Interference Suppression Techniques In Spread Spectrum Communications," *Proc. IEEE 1988 Annual Workshop on Information Theory,* pp. 6-16, April 1988.

[7] L. D. Davisson, "The Prediction Error of Stationary Gaussian Time Series of Unknown Covariance," *IEEE Trans. Inform. Theory,* vol. IT-19, no. 4, pp. 527-532, Oct. 1965.

[8] L. D. Davisson, "The Adaptive Prediction of Time Series," in *Proc. Nat. Electronics Conf.*, vol. 22, pp. 557-561, 1966.

[9] H. Akaike, "Fitting Autoregressive Models For Prediction," *Ann. Inst. Statist. Math.*, vol. 21, no. 2, pp. 243-247, 1969.

[10] H. Akaike, "Statistical Predictor Identification," *Ann. Inst. Statist. Math.*, vol. 22, no. 2, pp. 203-217, 1970.

[11] H. Akaike, "Information Theory and An Extension of the Maximum Likelihood Principle," in *Proc. 2nd Int. Symp. Information Theory,* pp. 267-281, 1972.

[12] H. Akaike, "Use of An Information Theoretic Quantity for Statistical Model Identification," in *Proc. 5th Hawaii Int. Conf. System Sciences,* pp. 249-250, 1972.

[13] H. Akaike, "A New Look At the Statistical Model Identification," *IEEE Trans. Automat. Contr.*, vol. AC-19, no. 6, pp. 716-723, Dec. 1974.

[14] H. Akaike, "A Bayesian Analysis of the Minimum AIC Procedure," *Ann. Inst. Statist. Math.*, vol. 30, part A, pp. 9-14, 1978.

[15] E. Parzen, "Some Recent Advances in Time Series Modeling," *IEEE Trans. Automat. Contr.*, vol. AC-19, no. 6, pp. 723-730, Dec. 1974.

[16] E. Parzen, "Multiple Time Series: Determining the Order of Approximating Autoregressive Schemes," *Multivariate Analysis - IV,* ed. by P. Krishnaiah, North Holland: Amsterdam, pp. 283-295, 1977.

[17] R. H. Jones, "Identification and Autoregressive Spectrum Estimation," *IEEE Trans. Automat. Contr.*, vol. AC-19, no. 6, pp. 894-898, Dec. 1974.

[18] R. H. Jones, "Autoregression Order Selection," *Geophys.*, vol. 41, pp. 771-773, Aug. 1976.

[19] G. Schwartz, "Estimating the Dimension of a Model," *Ann. Statist.*, vol. 6, pp. 461-464, 1978.

[20] J. Rissanen, "Modeling by Shortest Data Description," *Automatica,* vol. 14, pp. 465-471, 1978.

[21] J. Rissanen, "A Universal Prior for Integers and Estimation by Minimum Description Length," *Ann. Statist.*, vol. 11, no. 2, pp. 416-431, June 1983.

[22] J. Rissanen, "Universal Coding, Information, Prediction, and Estimation," *IEEE Trans. Inform. Theory,* vol. IT-30, no. 4, pp. 629-636, July 1984.

[23] E. J. Hannan and B. G. Quinn, "The Determination of the Order of an Autoregression," *Jour. Roy. Statist. Soc.*, Ser. B, vol. 41, no. 2, pp. 190-195, 1979.

[24] E. J. Hannan, "The Estimation of the Order of an ARMA Process," *Ann. Statist.*, vol. 8, no. 5, pp. 1071-1081, 1980.

[25] W. A. Fuller and D. P. Hasza, "Properties of Predictors for Autoregressive Time Series," *Jour. Am. Statist. Assoc.*, vol. 76, no. 373, pp. 155-161, Mar. 1981.

[26] R. J. Bhansali and D. Y. Downham, "Some Properties of an Autoregressive Model Selected by a Generalization of Akaike's FPE Criterion," *Biometrika,* vol. 64, no. 3, pp. 547-551, 1977.

[27] R. J. Bhansali, "Effects of Not Knowing the Order of an Autoregressive Process on the Mean Squared Error of Prediction - I," *Jour. Am. Statist. Assoc.*, vol. 76, no. 375, pp. 588-597, Sept. 1981.

[28] R. R. Bitmead, "Convergence in Distribution of LMS-Type Adaptive Parameter Estimates," *IEEE Trans. Automat. Contr.*, vol. AC-28, no. 1, Jan. 1983.

[29] R. R. Bitmead, "Convergence Properties of LMS Adaptive Estimators with Unbounded Dependent Inputs," *IEEE Trans. Automat. Contr.*, vol. AC-29, no. 5, May 1984.

[30] L. Gyorfi, "Adaptive Linear Procedures Under General Conditions," *IEEE Trans. Inform. Theory,* vol. IT-30, no. 2, pp. 262-267, Mar. 1984.

[31] N. Kunitomo and T. Yamamoto, "Properties of Predictors in Misspecified Autoregressive Time Series Models," *Jour. Am. Statist. Assoc.*, vol. 80, no. 392, pp. 941-950, Dec. 1985.

[32] T. L. Lai and C. Z. Wei, "Extended Least Squares and Their Applications to Adaptive Control and Prediction in Linear Systems," *IEEE Trans. Automat. Contr.*, vol. AC-31, no. 10, pp. 898-906, Oct. 1986.

[33] T. C. Butash and L. D. Davisson, "An Overview of Adaptive Linear Minimum Mean Square Error Predictor Performance," in *Proc. 25th IEEE Conf. Decision and Control,* pp. 1472-1476, Dec. 1986.

[34] C. Z. Wei, "Adaptive Prediction by Least Squares Predictors in Stochastic Regression Models With Applications to Time Series," *The Ann. of Statistics,* vol. 15, no. 4, pp. 1667-1682, 1987.

[35] M. Wax, "Order Selection for AR Models by Predictive Least Squares," *IEEE Trans. Acoust., Speech, Signal Processing,* vol. 36, no. 4, pp. 581-588, April 1988.

[36] L. D. Davisson and T. C. Butash, "Adaptive Linear Prediction and Process Order Identification," in *Proc. IEEE 1988 Annual Workshop on Information Theory,* pp. 20-32, April 1988.

[37] T. C. Butash and L. D. Davisson, "On The Design and Performance of Adaptive LMMSE Predictors," in *Proc. 1988 IEEE Int. Symp. Information Theory,* June 1988.

[38] A. Krieger and E. Masry, "Convergence Analysis of Adaptive Linear Estimation for Dependent Stationary Processes," *IEEE Trans. Inform. Theory,* vol. IT-34, no. 4, pp. 642-654, July 1988.

[39] E. J. Hannan, "Rational Transfer Function Approximation," *Stat. Science,* vol. 2, no. 2, pp. 135-161, 1987.

[40] J. L. Doob, *Stochastic Processes,* John Wiley, New York, 1952.

[41] P. H. Diananda, "Some Probability Limit Theorems With Statistical Applications," *Proc. Cambridge Philos. Soc.,* vol. 49, pp. 239-246, Oct. 1952.

[42] H. Cramér, *Mathematical Methods of Statistics,* Princeton University Press, Princeton, New Jersey, 1946.

[43] M. Rosenblatt, "A Central Limit Theorem and A Strong Mixing Condition," *Proc. Nat. Acad. Sci.,* vol. 42, pp. 43-47, 1956.

[44] P. Hall and C.C. Heyde, *Martingale Limit Theory and its Application,* Academic Press, New York, 1980.

[45] R. M. Gray and L. D. Davisson, *Random Processes: A Mathematical Approach for Engineers,* Prentice-Hall, Englewood Cliffs, New Jersey, 1986.

[46] P. Billingsley, *Convergence of Probability Measures,* John Wiley & Sons, New York, 1968.

[47] M. Iosifescu and R. Theodorescu, *Random Processes and Learning,* Springer-Verlag, New York, 1969.

[48] A. N. Kolmogorov and Y. A. Rozanov, "On Strong Mixing Conditions for Stationary Gaussian Processes," *Theory Prob. Appl.,* vol. 5, pp. 204-208, 1960.

[49] Y. A. Rozanov, *Stationary Random Processes,* Holden-Day, San Francisco, California, 1967.

[50] I. A. Ibragimov and Yu. V. Linnik, *Independent and Stationary Sequences of Random Variables,* Wolters-Noordhoff, Groningen Netherland, 1971.

[51] M. Rosenblatt, *Random Processes,* Springer-Verlag, New York, 1974.

[52] T. C. Butash, "Adaptive Linear Prediction and Process Order Identification," Ph.D. Dissertation, University of Maryland at College Park, 1990.

LINEAR PREDICTION TECHNIQUES IN SPEECH CODING

A. Gersho
University of California, Santa Barbara, CA, USA

1. ABSTRACT

Recent advances in speech coding algorithms and techniques based on the use of linear prediction now permit high quality voice reproduction at remarkably low bit rates. This paper reviews some of the main ideas underlying the algorithms of major interest today. The concept of removing redundancy by linear prediction is reviewed, first in the context of predictive quantization or DPCM. Then linear predictive coding, adaptive predictive coding, and vector quantization are discussed. The concepts of excitation coding via analysis-by-synthesis linear predictive coding is explained and some important enhancements such as vector sum excitations, and adaptive postfiltering are described. Low-delay coding by backward computation of LPC parameters is explained. The concept of phonetic segmentation of speech for closed-loop coding systems is also presented. Linear prediction is the key technique that underlies almost all of the important algorithms for speech coding of interest today. Finally, we discuss some recent work on nonlinear prediction of speech and its potential for the future of speech coding.

2. INTRODUCTION

With the advent of rapidly increasing digital signal processing technology, it has recently become cost effective to use rather sophisticated speech coding algorithms in numerous commercial, government, and military communications applications. Speech coding is already being or becoming widely used in many storage applications where the communication process is not necessarily to transport voice from one geographical location to another but from one point in time to a later point in time.

In this paper, we first describe some of the basic concepts of speech coding. Then we lead into the description of the main algorithms of interest today by starting with the basic ideas of predictive quantization, DPCM, LPC vocoders, and APC coders. We next introduce the idea of vector quantization, then come to excitation coding and coders based on analysis-by-synthesis coding and focus particularly on CELP or VXC type coders. Some recent developments of importance, vector sum excitation codebooks, low-delay VXC, and adaptive postfiltering are reviewed. Following this we introduce the use of phonetic segmentation in speech coding, a new approach that may contribute to the next generation of speech coders.

3. BASIC CONCEPTS OF SPEECH CODING

The signals shown in Fig. 1 illustrate the great variety in the character of speech waveforms. Sometimes periodic or almost periodic, other times a mixture of periodic and random-like signals and sometimes the waveform appears like random noise. Shown in the figure is a 10 ms time interval. A speech coder operating, for example, at 4 kb/s must be able to describe any such 10 ms segment (80 samples) using only 40 binary digits in such a way that the segment will be reproduced with an accuracy sufficient to insure that it will sound very close to the original. Unlike PCM where 8 bits are used to code each sample, in such a low bit rate coder we have only 1/2 of a bit available per sample to describe the sound or the waveform. Of course there is no way to adequately describe the amplitude of a sample, even if an entire bit were available per sample (as in the case of an 8 kb/s coder). Thus we must use clever techniques to exploit redundancy across samples by introducing memory in the encoding process, so that we do not merely examine one sample at a time and code that sample (as in PCM), but we store up past samples, and/or information obtained from past samples to send out essential digital information that will help us to specify the current sample. At the heart of almost all speech coders is the use of linear prediction as a redundancy removal method.

Speech coders have been traditionally grouped into vocoders (from "voice coders") and waveform coders. Today this dichotomy has become blurred with the current generation of so-called hybrid coders which embody some aspects of both of the above categories. Hybrid coders do attempt to reproduce the waveform, to some degree, while also describing key parameters that help to reproduce (synthesize) a sound perceptually similar to the original. Most of these coders make strong use of linear prediction methods for modeling the speech signal.

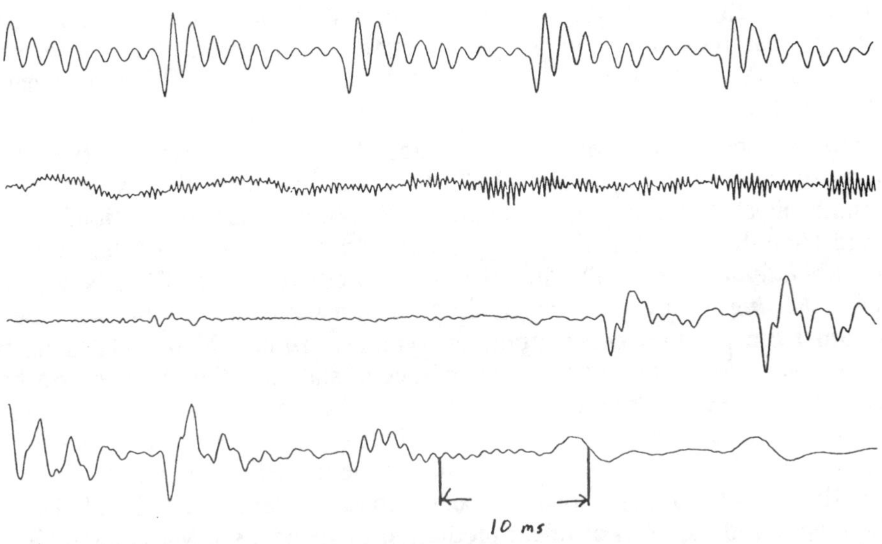

Fig. 1 Examples of Speech Waveforms

We assume the reader is familiar with PCM which, as used in telephony today, samples voice at 8,000 samples/s and codes each sample with an 8 bit word using a nonuniform quantizer based roughly on a logarithmic companding characteristic. Nothing further will be mentioned about this. Suffice it to note that a quantizer can be viewed as the cascade of an encoder (A/D converter) and a decoder (D/A converter). The encoder generates an index as a binary word specifying the amplitude level of the quantized value which approximates the input amplitude. Often the quantizer is viewed as a black box that generates both the index and the quantized level. The decoder (D/A) sometimes is called an 'inverse' quantizer and it simply maps the index into the reproduced level.

4. LINEAR PREDICTION

We assume the readers are familiar with the basic ideas of linear prediction for a stationary random process. In speech coding, the waveform is regarded as locally stationary with a short term spectral character that typically remains constant within an interval of 20 ms but can vary substantionally over much longer intervals of time. Speech is typically divided into fixed intervals called *frames* where a frame is typically of the order of magnitude of 20 ms long. Instead of using a single time-invariant predictor for linear prediction of speech, a new predictor is typically designed for each frame by buffering the frame of samples prior to coding and computing an mth order linear predictor that is optimal for the speech data in the current frame. many alternative and equivalent parameter sets can be used for indirectly characterizing the linear predictor coefficients. Most of these alternatives have the advantage that they are more robust to quantization errors than the coefficients themselves. A typical value of the predictor order m is 10 although occasionally larger values are used.

The prediction error filter which maps the input process into the prediction error (also called the *residual*) is an mth order FIR filter that is known to have minimum phase because of the properties of optimal linear prediction. If the order is large enough, the residual is approximately white so the the filter is sometimes called a *whitening filter*. The inverse of the prediction error filter is stable and is called a *shaping* filter or a *synthesis* filter since it recreates a signal with the speech spectrum from a white noise input. A synthesis filter which updates its transfer function every frame to suit the current speech statistics then serves as a basis for synthesizing a waveform with time-varying spectrum that is very similar to that of the original speech signal. Long-term prediction that makes use of the periodicity of voiced speech can also be used to enhance the modeling or synthesis process and is usually included in sophisticated speech coders today. For a detailed discussion of the theory and methods of linear prediction of speech, see Markel and Gray[1].

5. PREDICTIVE QUANTIZATION

A major advance in waveform coding of speech was the introduction of predictive quantization. Fig 2 shows the basic idea of this scheme.
First, note that a quantity $\hat{X}_k$ is subtracted from the the input sample X_k forming a difference sample d_k. This difference is quantized and then the quantity $\hat{X}_k$ is added back to the quantized approximation of the difference sample d_k, producing a final output $\tilde{X}_k$. Without giving any attention to how $\hat{X}_k$ is generated, it is evident that the error in approximating the input sample X_k by $\tilde{X}_k$ is exactly equal to the error incurred by the quantizer in approximating the difference signal. This means that if we can somehow make $\hat{X}_k$ very close to X_k, the difference signal will be small, and fewer bits will be needed for quantizing d_k so as to make the overall error in approximating X_k by $\tilde{X}_k$ also small. The quantity $\hat{X}_k$ is chosen to be a linear prediction of X_k based on previously reproduced samples. The predictor has transfer function

$$P(z) = \sum_{i=1}^{n} a_i z^{-i}$$

The difference between the input sample and its predicted value, (based on the past information known to the decoder), is quantized and the index specifying the quantized level of this difference is sent to the decoder. Note that the encoder contains a copy of the decoder.

The decoder replicates the feedback loop of the encoder. Note that the linear predictor now appears imbedded in a feedback loop. The decoder is simply an inverse quantizer which reproduces the sequence of quantized difference samples

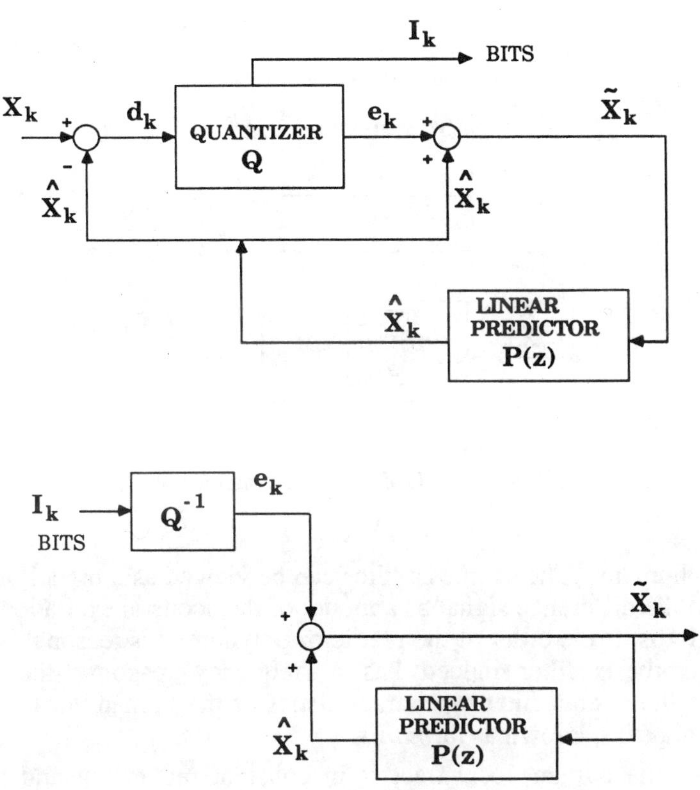

Fig. 2 Predictive Quantization

and feeds it into a filter with transfer function $[1-P(z)]^{-1}$, called the synthesis filter, to reproduce samples of the original signal X_k.

The performance gain of this structure is due to the prediction gain of the predictor, i.e. the ratio of variances of d_k to variance of X_k, or the factor by which the power of the input signal is reduced after removing the predictable error. This prediction gain in dB is what determines the performance improvement over straight PCM.

Fig. 3 shows the block diagram of a DPCM coder in a more conventional form, which is exactly the same scheme as in Fig. 2, only drawn in a less insightful way. By comparing the two figures, it is easily verified that they represent identical coders.

It is interesting to note that the DPCM decoder which generates speech from a sequence of difference samples models, in a primitive sense, the speech production

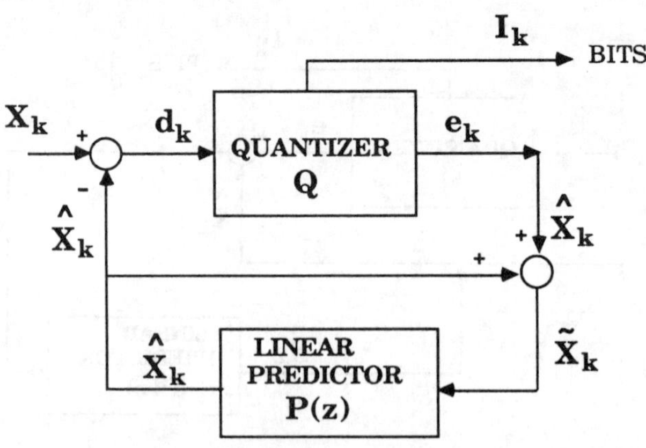

Fig. 3 Differential PCM in more conventional form

mechanism in humans. The synthesis filter can be viewed as a model of the human vocal tract and the difference signal as a model of the acoustic excitation signal produced at the glottis. If the order of the predictor polynomial is reasonably high, (8 or higher) the synthesis filter indeed has a frequency response that reasonably corresponds to the overall filtering characteristics of the human vocal tract with its distinct spectral peaks, known as *formants*.

Of course, the human vocal tract is in constant movement and thus its frequency response varies substantially in time, from one phonetic sound unit, or phoneme, to another. Only over a time interval of a few milliseconds is it likely to be more or less constant. In Adaptive DPCM (ADPCM), the predictor is also time varying and thereby has a greater ability to model the speech production mechanism.

Another improvement in DPCM is the use of pole-zero prediction. Fig. 4 shows the predictive quantization structure modified by the use of two predictors $P_1(z)$ and $P_2(z)$. Each takes a linear combination of past values from its input. The new predictor, P_2, is applied directly to the quantized difference samples, while P_1 combines these with the preceding value of $\hat{X}_k$, to produce the current value of $\hat{X}_k$. Note that the corresponding decoder structure, also shown in Fig 4, has a pole-zero synthesis filter, where P_2 contributes zeros and P_1 poles to the synthesis filter.

Indeed, the pole-zero filter may also provide a more versatile model of the human vocal tract if indeed a suitable number of poles and zeros were used and if the synthesis filter is adaptive (and thus time varying) to track the changing shape of the vocal tract. The CCITT 32 kb/s ADPCM standard, based on this structure, has 6 zeros and 2 poles and performs backward adaptation to make the two predictors

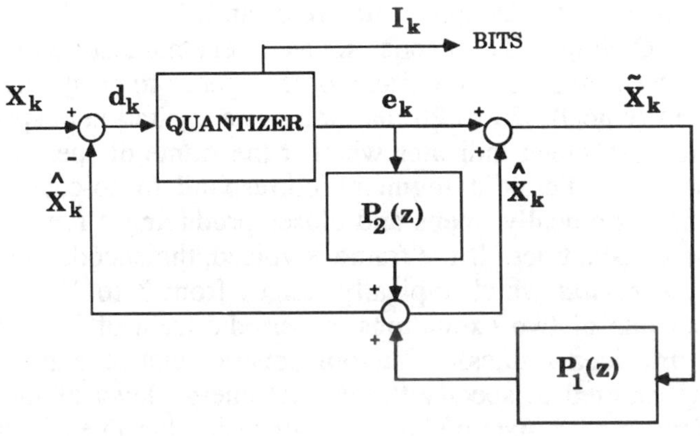

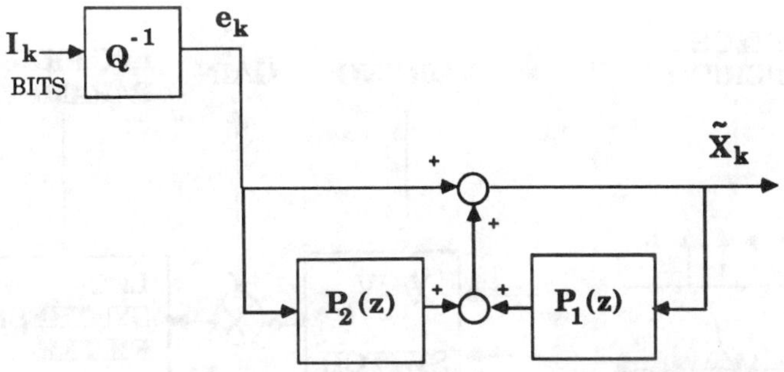

Fig. 4 Predictive Quantization (DPCM) with Pole-Zero Prediction

track the time-varying statistics of the speech.

6. LPC VOCODER

In an entirely different approach to speech coding, known as parameter-coding, analysis/synthesis coding, or vocoding, no attempt is made at reproducing the exact speech waveform at the receiver, only a signal perceptually equivalent to it. Early versions of this approach included formant synthesizers and so-called "terminal analog synthesizers". However, the most widely used form today was partly motivated by recognizing the DPCM decoder as a model of the speech production mechanism. The idea is to replace the quantized difference signal by a simple excitation signal which at least crudely mimics typical excitation signals generated in the human glottis.

Figure 5 illustrates the decoder structure of an LPC Vocoder. (LPC stands for Linear Predictive Coding.) The encoder sends a very modest number of bits to the decoder to describe each successive frame of the speech to be synthesized. A frame is a time segment typically 20 to 25 ms long. The excitation is specified by a one bit voicing parameter which indicates whether the frame of speech is judged to be periodic or aperiodic. Periodic segments correspond to so-called *voiced* speech where the glottis periodically opens and closes producing a fairly regular train of *pitch* pulses to the vocal tract. If the frame is voiced, the encoder also sends an estimate of the *pitch period* which typically ranges from 3 to 18 ms. The decoder locally generates one of two excitations, a periodic train of impulses at the pitch period, or (for unvoiced frames) a random noise excitation signal. A gain value must also be transmitted to specify the correct energy level of the current frame. Thus the set of parameters specified for the synthesis filter in each frames are: voicing decision, pitch (if appropriate), LPC coefficients (typically 10) and gain.

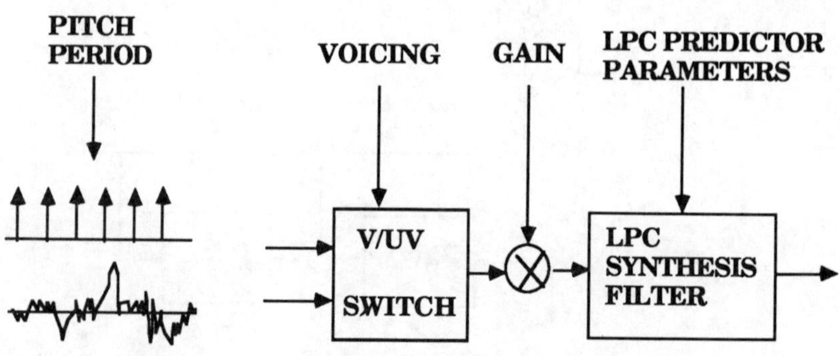

Fig. 5 LPC Vocoder – Decoder

The encoder of an LPC vocoder, also shown in Fig. 6, performs computations on each frame of input speech to determine the set of parameters needed by the decoder.

The linear predictor described here and in the context of DPCM is often called a *short-term* predictor or *formant* predictor. For later convenience we denote the short-term predictor by $P_s(z)$ where s indicates short. These names illustrate the fact that the predictor exploits the short-term correlation in nearby samples of the speech waveform, and the fact that it is the short-term correlation which characterizes the formants dominating the envelope of the speech spectrum. Generally three or four principal formants are evident in examining the Fourier transform of a speech frame. The formant synthesis filter has a frequency response whose magnitude closely corresponds to the envelope of the spectrum. The transfer function of this synthesis filter is $[1 - P_s(z)]^{-1}$.

Note that the vocoder scheme does not actually attempt to encode the speech waveform but only extracts some parameters or features that partially characterize each frame. Thus it does not have the ability to reproduce an approximation to the original waveform. Nevertheless, it can synthesize clear, intelligible speech at the very low bit-rate of 2400 b/s. Such vocoders have served for years as the underlying technology for secure voice terminals, which include the functions of encrypting a bit stream and digital modulation into an analog voiceband signal suitable for transmission over an analog telephone connection.

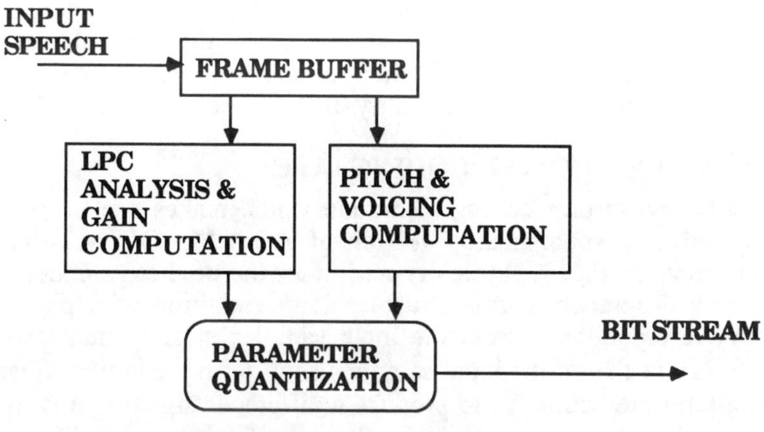

Fig. 6 LPC Vocoder – Encoder

7. PITCH PREDICTION

Another fundamental technique that has had a major impact on speech coding is the use of long-term or "pitch" prediction. The periodic, or nearly-periodic character of a speech segment suggests that there is considerable redundancy that can be exploited by predicting current samples from samples observed one period earlier. Because this periodicity is closely associated with the so-called fundamental frequency or *pitch* of voiced speech, the number of glottal openings per second, the repetition period is often called the *pitch period.* A long-term predictor or "pitch predictor" can be directly used to remove the periodicity when the period is known. The phrase "long-term" refers to the relatively large delay (many samples) used in pitch prediction compared to the small values for the short-term predictor. Thus, a pitch predictor typically has the transfer function

$$P_L(z) = \sum_{j=-i}^{i} \alpha_i z^{-m-i}$$

where m is the pitch period measured in samples, i is a small integer, and α_i are coefficients. Often a single tap predictor is used so that $i = 0$. The filter structure with transfer function $1 - P_L(z)$ removes periodicity, and thereby redundancy, by subtracting the predicted value from the current sample. This gives rise to a pitch synthesis filter, with the inverse transfer function $[1-P_L(z)]^{-1}$ which introduces a periodic character to an aperiodic input. We shall see how the pitch synthesis filter or *long-term* synthesis filter will play an important role in the new generation of speech coders.

The computation of the pitch predictor parameters, i.e. the pitch period and predictor coefficients, can be performed by the encoder in a manner similar to that used for LPC analysis where the buffered input speech is used to compute the predictor parameters. This is called an *open-loop* pitch analysis in contrast with a more recent method, to be described later, which optimizes the pitch predictor by directly measuring its impact on the overall quality of the speech reproduced by the decoder.

8. ADAPTIVE PREDICTIVE CODING (APC)

The oldest waveform coding technique which makes use of pitch prediction can be viewed as a sophisticated version of ADPCM. One version of an APC encoder is shown in Fig. 7. It clearly resembles the predictive coder of Fig. 2. In fact, the main difference in this structure is the addition of a pitch predictor to further remove redundancy from the input samples prior to quantization. In this scheme, we subtract from the input sample a short term prediction $\hat{X}_k$ and then subtract a long-term prediction $\hat{X}_k$ to produce a difference signal d_k that has very little redundancy compared to the original sequence of speech samples X_k. Note that with this structure, the exact same prediction values are added back to the quantized difference signal e_k so that we have, as in DPCM, the property that the overall error between the original speech and the reconstructed speech $\tilde{X}_k$ is equal to the quantization error $d_k - e_k$.

A crucial distinction between APC and DPCM, not indicated in the figure, is that the short- and long-term predictors are updated for every frame, by directly computing the necessary parameters from a frame of speech stored in an input buffer prior to being encoded. This implies that side information describing the predictor parameters must be multiplexed with the bits produced by the quantizer to specify the difference signal often called the prediction *residual* signal. In fact, in typical APC coders a rather low bit-rate is found to be adequate to code the residual signal

The decoder for this APC scheme is also shown in Fig. 7 and it is evident that it reproduces the same sample sequence $\tilde{X}_k$ as generated in the encoder.

What is most noteworthy about the decoder structure is that the speech is being regenerated or *synthesized* by applying a signal e_k to a cascade of two synthesis

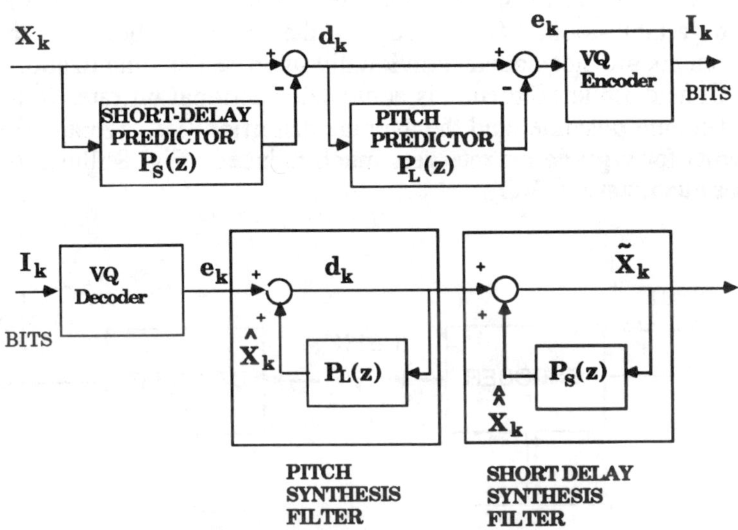

Fig. 7 Adaptive Predictive Coder

filters. If a reasonably good job was done in determining prediction parameters and updating them at a reasonably frequent rate, e.g., a frame rate of 20 ms, it is found that this signal is very closely described as white Gaussian noise. Thus in effect, we are synthesizing speech from a time-varying speech production filter by applying to it a particular white noise excitation. This paradigm will recur again in subsequent discussions.

various enhancements of APC have been developed, and in particular, quantization of the residual combined with entropy coding is often used. The APC structure can be modified by interchanging the role of long- and short-term prediction. APC speech coders have been implemented and used in the 1970s at typical bit rates of 9.6 kb/s and 16 kb/s. In the past decade, however, APC has gradually diminished in interest due to the emergence of newer and more powerful speech coding methods.

9. VECTOR QUANTIZATION

It has become recognized in the past decade that the efficient coding of a vector, an ordered set of signal samples or parameter values describing a signal, can be achieved by pre-storing a codebook of predesigned code vectors. For a given input vector, the encoder then simply identifies the address, or index, of the best matching code vector. Note that this is in essence a pattern matching algorithm. The index, as a binary word, is then transmitted and the decoder replicates the corresponding code vector by a table-lookup from a copy of the same codebook. In this way, the vector components are not coded individually as in scalar quantization, but rather all at once. Considerable efficiency is achieved, fractional bit rates (bits per vector component) become possible, and the average distortion (i.e., average squared error per component) for a given bit rate gets much reduced. Fig. 8 illustrates the basic idea of vector quantization (VQ).

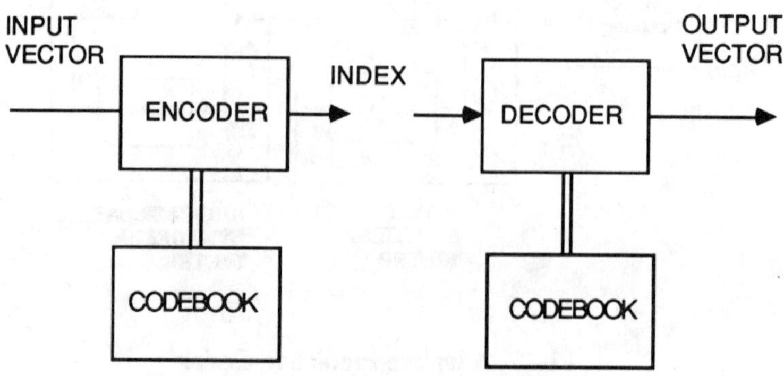

Fig. 8 Vector Quantization

The first major application of VQ to speech coding was reported by [2] where the bit rate of an LPC vocoder was substantially reduced by applying VQ to the LPC parameters. Subsequently VQ found its way into waveform coding as well and in particular a generalization of DPCM using vector prediction together with VQ was reported in [3]. Today VQ is a well-established and widely used technique. It has been applied to the efficient coding of the LPC parameter set, the pitch predictor filter parameters, as well as to Vector PCM (VPCM), the coding of a waveform by partitioning it into consecutive blocks (vectors) of samples.

10. OPEN LOOP VECTOR PREDICTIVE CODING

To illustrate the use of VQ, let us return to the APC scheme described above and consider that the largest contribution to the bit-rate of an APC coder is the coding of the residual waveform. However, in the structure of Fig. 7 the residual is generated only one sample at a time, and the next residual sample depends on feeding back the previous sample for obtaining the next short-term prediction. Thus the structure is not immediately amenable to VQ which requires storing up a block of residual samples before performing the pattern matching operation. There are two ways to circumvent this obstacle. One is based on a vector generalization of ADPCM introduced in [3] and extended to a vector version of APC in [4]. The other is simply to modify the encoder structure by removing the feedback around the quantizer, and generate the prediction residual by an open-loop method as is shown in Fig. 9. Note that the decoder has the same synthesis filter structure as that of the more conventional APC scheme. Here VPCM is applied to the residual signal, and since many of its samples may be encoded by a few bits, fractional bit rates (i.e. less than 1 bit per sample) can be attained.

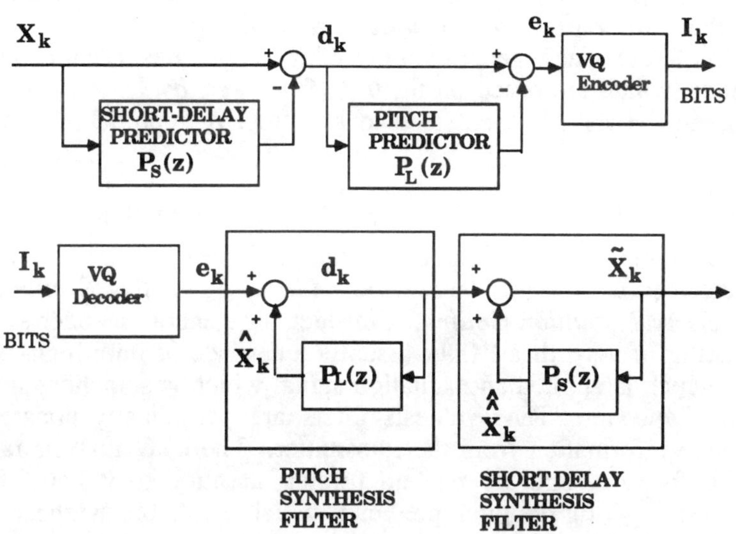

Fig. 9 Residual Encoding with Vector Quantization

Although this scheme has been applied by several researchers to speech coding, it suffers from one major disadvantage. Unlike the previous APC scheme, the overall error between original and reproduced speech in this coder is not equal to the error produced by the quantizer. Ordinarily, a VQ codebook is optimally designed to minimize the average distortion between input and reproduced vectors and encoding is performed by simply selecting the code vector best matching given input vector. In this coding scheme this implies that the reproduced residual is made to approximate the unquantized residual as closely as possible. However, this is *not* an optimal strategy, since our objective is to make the reproduced *speech* as close as possible to the original speech. With the predictor filters time-varying, these turn out not to be identical criteria, as the relationship between the error in quantizing the residual to the error in reproducing the original speech is a very complex one and varies from frame to frame.

These observations suggest that regardless of whether we use scalar or vector quantization or any other mechanism for digitally specifying an excitation signal for the decoder, the main task for the encoder is to figure out what excitation will do the best job of reproducing the original speech. The encoder structure of Fig. 9 incorporates a somewhat *ad hoc* mechanism for selecting an excitation vector from the codebook, which focuses narrowly on the residual signal, rather than on the speech itself. This is an intrinsic limitation of the open loop structure.

Let us therefore discard this encoder, and consider what is the best possible structure that can be used to supply data to the decoder given in Fig. 9. This perspective has led to a new generation of coding techniques, often called *hybrid* coding methods, which are based on the use of *analysis-by-synthesis* to determine the best excitation signal that will lead to an effective reproduction of the original speech.

11. ANALYSIS-BY-SYNTHESIS EXCITATION CODING

We now examine the most important family of speech coding algorithms known today, described as *Analysis-by-Synthesis Linear Predictive Excitation Coding* or more concisely *Excitation Coding*. Consider the general decoder structure of Fig. 10, consisting of a synthesis filter (usually a cascade of both long- and short-term filters) to which is applied an excitation signal which is somehow specified by bits sent by the encoder. The synthesis filters are periodically updated, usually by separate side information from the transmitter. The LPC analysis task is classical and straightforward, and we pay no further attention to it here. The open-loop method for computing the pitch predictor, which yields the synthesis filter parameters, was described earlier.

The encoder contains a copy of the decoder so that for any excitation waveform it can generate the same speech signal as the decoder would. Given a bit allocation and a mechanism for generating such waveforms, the encoder actually generates by trial and error all possible excitation signals for each time segment. The key idea here is that we try a large family of possible excitation segments and then apply each member in turn to the synthesis filter (the speech production model). For each synthesized segment we can compute a quantitative distortion measure, which indicates how badly the segment differs from the intended original. This process is repeated until the best excitation segment is found. Then, and only then, is the binary word specifying the best excitation segment transmitted.

The task of finding an appropriate excitation signal copying the decoder at the encoder, can be viewed as an *analysis* process, since in some sense we are extracting an appropriate excitation signal from the original speech. The method is called *analysis-by-synthesis* because this is done by synthesizing the speech segment that each candidate excitation would produce to examine how well it reproduces the original speech.

There are three principal mechanisms for generating excitation signals for this class of coding systems, known as tree or trellis coding, multipulse coding, and VQ.

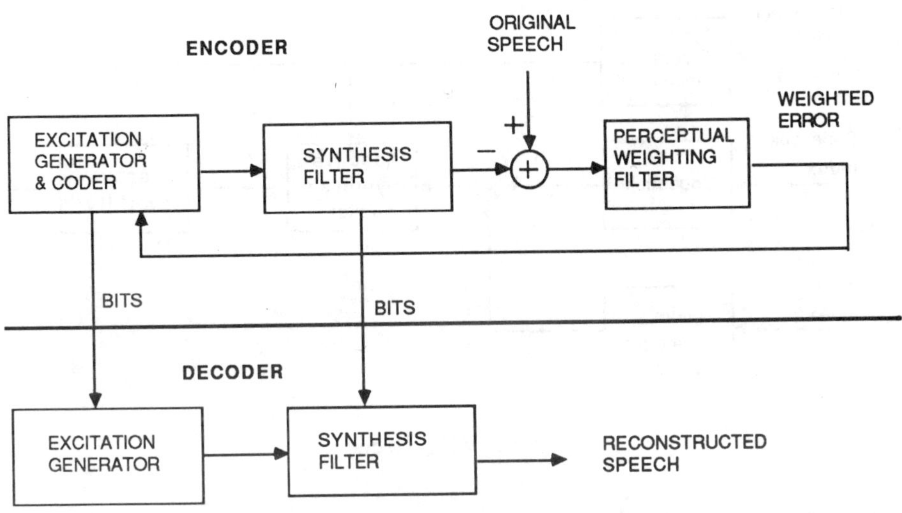

Fig. 10 Excitation Coding

While all three are of interest, the third is most widely used, and we focus on this approach in the sequel. The generic coding algorithm for the use of a VQ codebook is called Vector Excitation Coding (VXC), also known as Code-Excited Linear Prediction (CELP). This has led to many powerful speech coders for bit rates ranging from 4.8 to 16 kb/s.

12. VECTOR EXCITATION CODING

A generic VXC decoder structure is shown in Fig. 11. It is natural to describe the decoder first since it determines how the speech can be synthesized from transmitted data. Then encoder is in a sense a servant of the decoder, since its job is to examine the original speech and determine the best data to supply the decoder. The decoder receives and demultiplexes the data needed to specify the synthesis filter parameters, the excitation code vector, and in addition, a gain-scaling factor. A standard technique in VQ is to take advantage of the fact that owing to the wide dynamic range of speech, similarly shaped waveform portions may occur with different amplitudes, so that one may attribute to each segment a "gain" and a "shape" property. These attributes can then be handled separately via different codebooks, avoiding the ineffient duplication of waveform segments of similar shape, differing only in energy. By this method both codebook size and search complexity can be reduced.

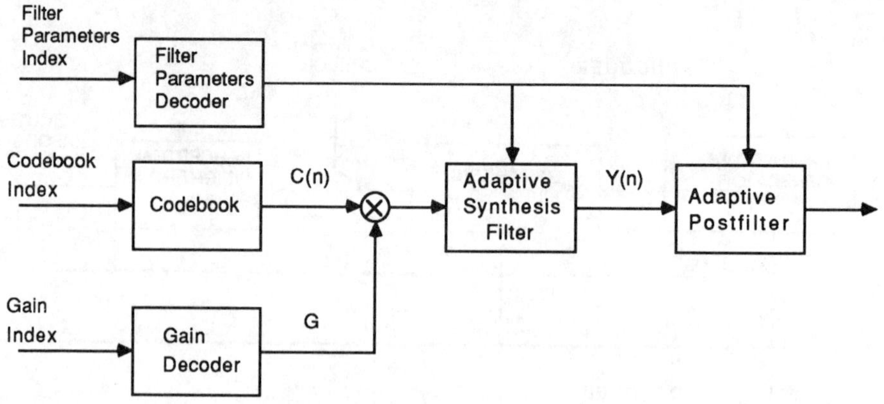

Fig. 11 VXC Decoder

It has been found empirically that the parameters of the synthesis filter need to be updated less frequently than new excitation vectors need to be supplied. For a 4.8 kb/s coder a typical frame size, i.e. the time span between successive updates of the synthesis filter, is 20-30 ms, while the excitation vector dimension, called *subframe*, may be a quarter of this. For higher bit-rate coders there may be even more subframes in a frame.

For each subframe, the decoder receives a sequence of $c + g$ excitation code bits which identify a pair of indexes which specify one of 2^c excitation code vectors and one of 2^g gain levels, both by means of a table-lookup procedure. This leads to a gain-scaled excitation vector with dimension k. This vector is serialized as k successive samples and is applied to the synthesis filter. The filter is clocked for k samples, feeding out the next k samples of the synthesized speech; then it is "frozen" until the next scaled excitation vector is available as the next input segment to the synthesis filter.

In many applications an adaptive postfilter is added to the decoder as a final postprocessing stage, to enhance the quality of the recovered speech. This filter is adapted to correspond to the short term spectrum of the speech. We shall later describe the operation of the adaptive postfilter; however, for now we ignore it since it is not a fundamental or essential component of VXC.

13. The VXC Encoder

The VXC encoder structure is shown in Fig. 12. We describe its operation in the simplest way, while ignoring the many short cuts and tricks which greatly reduce the complexity involved in the search process. The encoder receives input speech samples which are grouped into blocks of k contiguous samples, each regarded as a vector. On the arrival of each such vector, the task of the encoder is to determine the next $c+g$ bits of data to be transmitted to the decoder so that the decoder will then be able to synthesize a reconstructed output speech vector that closely approximates the original input speech vector.

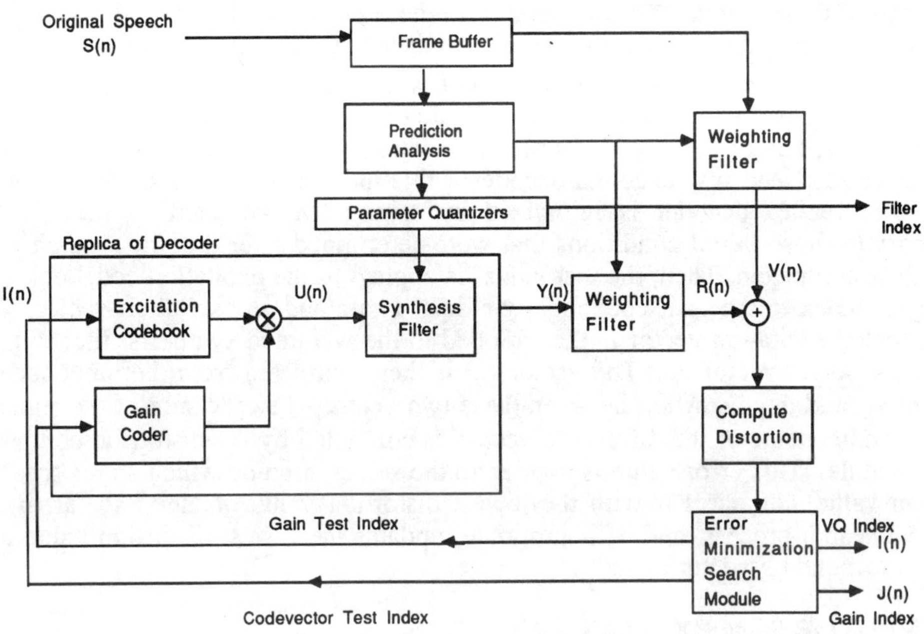

Fig. 12 VXC Encoder

This implies that the encoder embody a replica of the decoder, as shown in Fig. 12, which can locally generate each of the $N = 2^{c+g}$ possible speech vector candidates that the decoder would produce for the same transmitted data values. However, the replica decoder does not include the postfilter used in the actual decoder.

In order to search for a reproduction that is closest, in a perceptually meaningful sense, to the original speech, a *perceptual weighting filter* is used to modify both the original input speech and the reconstructed output speech vector before the distortion between the two is measured. Note that the weighting filter is combined with the synthesis filter to give a weighted synthesis filter with a modified transfer function that is distinct from the synthesis filter used in the decoder. Speech samples emerging from the weighting filter are also configured into corresponding vectors of k contiguous samples, called "weighted speech vectors"

Since the replica decoder is operating repeatedly in the search process, we must ensure that each candidate output speech vector, corresponding to a candidate data index pair being tested, is produced under the same conditions as will be present when the actual decoder generates the next output vector. After each test of a candidate index pair, the memory state of the replica decoder has changed and is no longer at the correct initial condition for the next test. Therefore, before generating each of these candidates, the memory in the replica decoder (including the perceptual weighting filter) must also be reset to the correct initial conditions.

The error minimization search module sequentially generates a pair of test indexes, corresponding to a particular pair of code vector and gain level. These are fed to the replica of the decoder which generates a synthesized speech vector that would be produced by the actual decoder if this index pair were actually transmitted. The replica decoder is initialized by setting the weighted synthesis filter memory to those initial conditions that were determined after the prior search process was completed. Then, the test index, is applied to the excitation codebook and the gain index to the gain codebook, yielding a gain and an excitation vector. The gain scaled excitation vector is then applied to the weighted synthesis filter to produce the output vector $\mathbf{r}_n$. The vector $\mathbf{r}_n$ is then subtracted from the input speech vector $\mathbf{v}_n$ and the distortion between these two vectors, i.e., the sum of the squares of the components of the difference vector, is computed by the distortion computation module. This error value is applied to the search module which stores the distortion value, compares it with the lowest distortion value obtained so far in the current search process, and, if appropriate, updates the lowest distortion value and the corresponding vector index.

14. VECTOR SUM EXCITATIONS

A question of practical importance, is how the quality of a given VXC coder can be improved if more bits are made available and to which components of the coder should these bits be assigned for the maximum benefit. It is generally recognized, that the best performance gain comes from increasing the codebook size. However adding just one bit per code vector doubles the codebook size and the corresponding search complexity. Thus, computational constraints of the available signal processor quickly force one to limit the codebook size and lead to alternative designs where the vector dimension is reduced and more bits are given to synthesis filter parameters. The use of specially constrained codebook structures offers the possibility of larger codebooks and significant performance improvements while maintaining tolerable complexity.

Gerson and Jasiuk recently introduced technique for reducing the complexity of the excitation codebook search procedure[5]. Rather than have each of M code vectors be independently generated either randomly or by a design procedure, they design b *basis* vectors and then generate the $M = 2^b$ code vectors by taking binary linear combinations of the basis vectors. The resulting coding algorithm, a derivative of VXC, is called Vector Sum Excited Linear Prediction (VSELP) and an 8 kb/s version of this algorithm has been adopted as a standard for the U.S. cellular mobile telephone industry. We next explain the basic idea of this technique for fast codebook search.

Let $\mathbf{v}_i$ denote the ith basis vector of a given set of b basis vectors. The code vectors are then formed as

$$\mathbf{u}_\theta = \sum_{i=1}^{b} \theta_i \mathbf{v}_i$$

by taking all possible linear combinations where $\theta_i = \pm 1$ for each i. Thus each binary-valued vector θ determines a particular code vector $\mathbf{u}_\theta$. Naturally, the b bit binary word transmitted over the channel can simply correspond to a mapping of θ values with +1 being a binary 1 and −1 being a binary 0. Since the code vectors are so simply generated, b basis vectors need be stored rather than storing an entire codebook of M code vectors.

This special codebook structure can be searched very efficiently. Instead of finding the vector output of the weighted synthesis filter for each of the M codevectors, only the filtered output of the b basis vectors need be determined because from these any synthesized output can be readily obtained by addition. Furthermore the search for the optimal codevector becomes computationally simplified by noting that the mean-squared error between the weighted input vector and a filtered codevector depends in a simple manner on the values of θ_i. By ordering the b bit binary word in a Gray code, only one bit changes from one word to the next. This means that only a simple change is needed to compute the mean-squared error for the next candidate code vector from the previous candidate code vector.

The vector sum approach can be augmented by using multiple-stage VXC [6]. and joint optimization of the gains for each stage. The joint optimization becomes easy to implement with the vector sum codebooks [5].

15. CLOSED-LOOP PITCH SYNTHESIS FILTERING

An alternative and improved method of designing the long-term predictor (LTP) filter was first proposed for the multipulse excitation coder [7] and later applied to vector excitation coders [8] [9] [10]. proposed for multipulse excitation coding and subsequently applied to VXC. Although it is of higher complexity and requires a higher bit-rate, it does offer superior performance. Furthermore, when the closed-loop LTP is used, the size of the excitation codebook is reduced and hence the computational load is reduced. The pitch lag and predictor coefficients of a closed-loop LTP are chosen in such way that the mean square of the perceptually weighted reconstruction error vector is minimized.

For a one-tap LTP, the predictor parameters can be determined two steps: (a) find the pitch lag m (from a predefined range) that maximizes a quantity that is independent of the prediction coefficient, and (b) compute the prediction coefficient from a simple formula.

In the closed-loop LTP method, the pitch lag ordinarily has to be greater or equal to the speech vector dimension in order to obtain the previous LTP output vector. Hence, the vector dimension, which is also the adaptation interval of the LTP, needs to be reasonably small to handle short pitch periods. Decreasing the adaptation interval increases the bit rate needed to code the LTP parameters.

16. ADAPTIVE POSTFILTERING

As already discussed, the perceptual weighting filter is a valuable component of a VXC encoder since it exploits the masking effect in human hearing by removing quantization noise from exposed frequency regions where the signal energy is low, and "hiding" it under spectral peaks. At bit rates as low as 4.8 kb/s or less, however, the average noise level is quite high and thus it is not possible to simultaneously keep the noise below the masking threshold at spectral valleys as well as at formant frequencies. Since the formant peaks are more critical for perceptual quality, at low bit rates the weighting filter tends to protect these regions while tolerating noise above the threshold in the valleys. The technique of adaptive postfiltering attempts to rectify this by selectively attenuating the reproduced speech signal in the spectral valleys. This somewhat distorts the speech spectrum in the valleys but it also reduces the audible noise. Since a faithful reproduction of the spectral shape is perceptually much less important in the valleys than near formants, the overall effect is beneficial and leads to a notáble improvement in subjective speech quality.

A more primitive form of adaptive postfiltering to enhance performance was applied to ADPCM by Ramamoorthy and Jayant[11] and to APC by Yatsuzuka [12]. Recently, an improved version for adaptive postfiltering was found [13] which is effective for VXC (or CELP).

For a filter to attenuate the spectral valleys, it must adapt to the time-varying spectrum of the speech. The synthesis filter parameters provide the needed information to identify the location of these valleys and are thus used to periodically update the postfilter parameters. since the LPC spectrum of a voiced sound typically has tilts downward at about 6 dB per octave, the corresponding all-pole postfilter will also have such a tilt causing undesirable muffling of the sound. This can be overcome by augmenting the postfilter with zeros at the same or similar angles as the poles but with smaller radii. The idea is to generate a numerator transfer function that compensates for the smoothed spectral shape of the denominator. The overall transfer function used for the postfilter in[13] is a pole-zero transfer function, given by:

$$H(z) = (1 - \mu z^{-1}) \frac{1 - P_s(z/\beta)}{1 - P_s(z/\alpha)}$$

An all-pole filter $[1-P_s(z/\alpha)]^{-1}$ introduces an undesired spectral tilt effect. The frequency response of the pole zero postfilter reveals the spectral tilt (and associated muffling effect) to be substantially reduced. Since the transfer function of the postfilter changes with each speech frame, a time-varying gain is produced. To avoid this effect, an automatic gain control is used.

We can think of the reproduced speech coming into the postfilter as the sum of clean speech and quantizing noise. Although the postfilter is of course attenuating spectral valleys of both the speech and the noise, the distorting effect of the filter on the speech is negligible due to the low sensitivity of the ear to changes in the level of the spectral valleys. This has been verified by applying the original (uncoded) speech to the adaptive postfilter: the original and filtered speech sound essentially the same.

Though postfiltering clearly improves the performance of a single coder, when multiple stages of coding and decoding follow each other, the postfilter in each stage introduces a slight degradation that accumulates with the number of stages. postfiltering may thus not be desired for applications with tandeming.

Pole-zero adaptive postfiltering following the approach described above has been included in the U.S. digital cellular telephone standard for 8 kb/s speech coding and as an optional feature in the U.S. government standard for 4.8 kb/s speech coding. [14]. Both standards are derivatives of VXC.

17. BACKWARD ADAPTIVE VXC FOR LOW DELAY CODING

Vector Excitation Coding (VXC) combines techniques such as vector quantization, analysis-by-synthesis codebook searching, perceptual weighting, and linear predictive coding to successfully achieve good speech quality at low bit rates. However, one important aspect of coding has been ignored in the development of VXC or other conventional low bit rate excitation coding schemes; that is the *coding delay*. In fact, most existing speech coders with rates at or below 16 kbps require high delays in their operation, and cause various problems when they are applied to practical communication systems. In case of VXC, a large net coding delay, excluding computational delays, results from the use of buffering needed to perform the LPC and open-loop pitch analysis. Recently, new methods have been proposed to adapt synthesis filters without the high coding delay mentioned above while maintaining the quality of encoded speech.

With the conventional VXC scheme described above, the synthesis filter is adaptively updated every frame using what is sometimes called forward adaptation, the process of recomputing and updating the desired filter parameters from the input speech. The use of the forward adaptation has two disadvantages: it requires transmission of side information to the receiver to specify the filter parameters and it leads to a large encoding delay of at least one analysis frame due to the buffering of input speech samples. The input buffering and other processing typically result in a one-way codec delay of 50 to 60 ms. In certain applications in the telecommunications network environment, coding delays as low as 2 ms per codec are required.

Recently, the CCITT adopted a performance requirement of less than 5 ms delay with a desired objective of less than 2 ms for candidate 16 kbit/s speech coding algorithms to be considered for a new standard intended to achieve the same quality as the 32 kb/s ADPCM standard, G.721. Such a low delay is not feasible with the established coders that are based on forward adaptive prediction coding systems. Although the 32 kbit/s ADPCM algorithm, CCITT Recommendation G.721, satisfies the low delay requirement, it cannot give acceptable quality when the bit rate is reduced to 16 kbit/s.

An alternative solution is based on a recently proposed backward adaptation configuration. In a backward adaptive analysis-by-synthesis configuration, the parameters of the synthesis filter are not derived from the original speech signal, but computed by backward adaptation extracting information only from the sequence of transmitted codebook indices. Since both the encoder and decoder have access to the past reconstructed signal, side information is no longer needed for synthesis filters, and the low-delay requirement can be met with a suitable choice of vector dimension.

VXC incorporated with backward adaptation to satisfy the low-delay requirement is called Low-Delay VXC or Low-Delay CELP. Two approaches to backward adaptation are studied, and they are classified as *block* and *recursive*. In the block algorithms, the reconstructed signal and the corresponding gain-scaled excitation vectors are divided into blocks (frames), and the optimum parameters of the adaptive filter are determined independently within each block. In the recursive algorithms, the parameters are updated incrementally after each successive excitation and reconstructed vector are generated.

To achieve the low-delay requirement, two versions of LD-VXC were proposed to the CCITT. One uses a codebook of dimension 5 and a very high order block-adaptive short-term predictor computed by LPC analysis on the previously reproduced speech. The other has a codebook of dimension 4 and uses a recursive backward adaptation method for a pole-zero predictor and for a pitch predictor. With the standard sampling rate of 8 KHz, we are allowed to use a codebook of size 256 at 16 kbit/s. Simulation results show that LD-VXC achieves an SNR of about 20 dB with either block or recursive adaptation. Transmission errors were also taken into account in the design of LD-VXC. With the help of leaky factors and pseudo-gray coding, the performance of the coder only degrades slightly at 0.1% error rate, and intelligible speech is produced even at error rate as high as 1%. More details are reported in[15] and [16].

18. VXC WITH PHONETIC SEGMENTATION

Although VXC achieves fairly high-quality speech at 4.8 kbps, the performance achieved with current VXC based algorithms degrades rapidly as the bit-rate is reduced below 4.8 kbps, leaving a substantial gap between the natural voice quality of VXC at 4.8 kbps and the synthetic quality attainable at 2.4 kbps (or higher) with an LPC vocoder. An important future direction for speech coding is to find coding algorithms that will achieve at 4 kb/s and below the natural quality attainable today with the best versions of VXC. One of the motivations for this interest is the next generation of digital cellular telephones where it is expected that a bit rate in the neighborhood of 4 kb/s will be required in order to meet the increasing channel capacity objectives.

One research direction that we have been studying, Phonetically-Segmented Vector eXcitation Coding (PS-VXC)[17], appears to show promise and might lead to a speech coder operating at bit rates significantly below 4.8 kb/s yet with a quality comparable to current 4.8 kb/s coders.

In this method, speech is segmented into a sequence of contiguous variable-length segments constrained to be an integer multiple of a fixed unit length. The segments are classified into one of six phonetic categories. This provides the front-end to a bank of VXC coders that are individually tailored to the different categories.

The motivation for this work derives from the fact that phonetically distinct speech segments require different coding treatments for preserving what we call *phonetic integrity*. With phonetic segmentation, we can assign the wide variety of possible speech segments into a small number of phonetically distinct groups. In each group, different analysis methods and coding strategies can be used to emphasize the critical parameters corresponding to important perceptual cues. It also becomes easier to identify each individual coding problem in isolated phonetic groups and optimize a multi-mode coding algorithm to suit various phonetic categories.

Table 1 summarizes the segment classification and coding structures used for these classes by specifying salient features and coder parameters for each of the six categories. Table 2 lists the bit-allocation for each category in PS-VXC. The details of the coding algorithm and recent improvements are reported in [17] and [18].

The three main segment types, if coded individually, would yield rates as follows: unvoiced — 3 kb/s, unvoiced/onset pairs — 3.6 kb/s, voiced — 3.6 kb/s. For typical speech files, the average rate is 3.4 kb/s, which could be achieved as a fixed rate with buffering of the encoder output. Alternatively, a fixed rate of 3.6 kb/s is readily attainable with some padding of the bit stream.

Informal listening tests indicate that the quality at a fixed 3.6 kb/s rate is roughly comparable to that of conventional VXC at 4.8 kb/s. Nevertheless, there is room for considerable improvement in both the coding algorithm for particular segment categories and in the definition and number of the phonetic classes used in the segmentation process. An end-to-end coding delay of approximately 100 ms (including overhead) is anticipated.

19. NONLINEAR PREDICTION OF SPEECH

A new method for the nonlinear prediction of one dimensional signals was recently developed called Nonlinear Predictive Vector Quantization (NLPVQ). It involves the approximation of conditional expectations by using vector quantization and sample averaging. The main motivation for this research was the inherent limitation of linear prediction, which is the most commonly used technique in speech coding. Its simplicity and low computational complexity, as well as the availability of various adaptation techniques have made linear prediction an attractive method for speech data compression. However, linear prediction is known to be suboptimal for non-Gaussian signals such as speech, and does not make use of the higher-order statistics which may contain valuable information. Nonlinear prediction of speech in principle offers a way to enhance the redundancy-removing feature of linear prediction and provide a superior model of the speech production mechanism.

We developed a novel method for the nonlinear prediction of speech which does not require a parametric model of the predictor. When the constraint of linear estimation is abandoned, the best least-squares estimate of a random variable is its conditional expectation given the observed variables. Although in general this operation requires the knowledge of the joint pdf of the signal, VQ provides a way of approximating it using the training data directly. Instead of conditioning the desired random variable on past observables, we condition it on a quantized version of the past. As the quantizer resolution gets higher, the approximation will become more accurate. Hence, the estimate will approach, in an asymptotic sense,the optimal mean-square estimate. This is essentially a special case of Nonlinear Interpolative Vector Quantization (NIVQ) [19]. The observable vector is quantized to a codevector at the encoder. The encoder then generates an index which the decoder uses to address the predicted value of the next sample in a prediction table.

The method was tested for a known stationary signal with non-Gaussian statistics and nonlinear dependencies between adjacent samples. We found in this case that the prediction gain obtained with NLPVQ is remarkably close to the theoretical optimum. NLPVQ was then applied to voiced and unvoiced speech and tested for various prediction orders, codebook sizes, and training set sizes. In each case, the resulting prediction gains were compared with results obtained for linear predictors designed from the same training set. For unvoiced speech, the Segmental Prediction Gain (SEGPG) obtained with NLPVQ was found to be higher than that obtained by LP even for small codebook sizes. For voiced speech, the SEGPG of second-order NLPVQ exceeds that of the linear predictor at sufficiently large codebook sizes (12-14 bits).

The most dramatic result of our experiments was the effect of NLPVQ on the residual signal. From informal listening tests, we observed a major drop in the intelligibility of the prediction residual for NLPVQ compared to LP. Moreover, the spectrum of the nonlinear prediction residual is significantly flatter than that of the LP residual. NLPVQ appears to be successful in eliminating the higher-order harmonics of the fundamental pitch frequency, thus potentially reducing the need for long term prediction.

While these results motivate further study of nonlinear prediction, practical application to speech coding may be premature. There are several nontrivial problems with nonlinear prediction that must be solved before it can become a useful technique for analysis-by-synthesis coding systems. Additional effort is necessary in the the joint optimization of the encoder and decoder, and the adaptation of the nonlinear predictor to time-varying signal statistics. Nevertheless, it is an interesting new direction in speech coding that is certainly worthy of further study.

20. CONCLUDING REMARKS

In this overview, we have only touched the surface of the rich and active field of speech coding. We have described some of the main concepts that underly speech coding algorithms of current interest today. In particular, linear prediction for both short and long term, analysis-by-synthesis, vector quantization, perceptual weighting for noise shaping, adaptive postfiltering, closed-loop pitch analysis, and vector-sum codebook structures. No doubt in the next few years, there will be new advances that we cannot anticipate today.

The motivation for the continued activity in speech coding research is in large part due to the combination of two factors: the rapidly advancing technology of signal processor integrated circuits and the ever increasing demand for wireless mobile and portable voice communications. The technology permits increasingly complex and sophisticated signal processing algorithms to become implementable and cost effective. Mobile communications and the emerging wide scale cordless portable telephones will increasingly stress the limited radio spectrum that is already pushing researchers to provide lower bit-rate and higher quality speech coding with lower power consumption, increasingly miniaturized technology, and lower cost. The insatiable need for humans to communicate with one another will continue to drive speech coding research for years to come.

21. ACKNOWLEDGMENT

This work was supported in part by the National Science Foundation, the University of California MICRO program, Rockwell International Corporation, Bell Communications Research, and Bell-Northern Research.

References

[1] J. D. Markel and A. H. Gray, Jr., *Linear Prediction of Speech*, Springer-Verlag, New York, NY, 1976.

[2] A. Buzo, A. H. Gray, R. M. Gray, and J. D. Markel, "Speech Coding Based upon Vector Quantization," *IEEE Trans. Acoust., Speech, and Signal Processing*, vol. ASSP-28, no. 5, pp. 562-574, October 1980.

[3] V. Cuperman and A. Gersho, "Vector Predictive Coding of Speech at 16 kbits/s," *IEEE Transactions on Communications*, vol. COM-33, pp. 685-696,

July 1985.

[4] J. H. Chen and A. Gersho, "Vector Adaptive Predictive Coding of Speech at 9.6 kb/s," *Proc. IEEE Inter. Conference on Acoust., Speech, and Signal Processing*, pp. 1693-1696, Tokyo, Japan, April 1986.

[5] I. A. Gerson, M. A. Jasiuk, "Vector Sum Excited Linear Prediction," *IEEE Workshop on Speech Coding for Telecommunications*, Vancouver, September 1989.

[6] G. Davidson, A. Gersho, "Speech Waveforms," *Proc. Inter. Conf. Acoust., Speech, & Signal Processing*, pp. 163-166, April 1988.

[7] S. Singhal and B. S. Atal, "Improving Performance of Multi-Pulse LPC Coders at Low Rates," *Proc. IEEE Inter. Conf. Acoustics, Speech, and Signal Processing*, vol. 1, pp. 1.3.1-1.3.4, San Diego, March 1984.

[8] R. C. Ross and T. P. Barnwell, "The Self-Excited Vocoder," *Proceedings of IEEE International Conference on Acoustics, Speech, and Signal Processing*, vol. 1, pp. 453-456, Japan, April, 1986.

[9] P. Kabal, J.L. Moncet, and C.C. Chu, "Synthesis Filter Optimization and Coding: Applications to CELP," *Proc.IEEE Inter. Conf. Acoust., Speech, and Signal Processing*, vol. 1, pp. 147-150, New York City, April 1988.

[10] W. B. Kleijn, D. J. Krasinski, R. H. Ketchum, and Improved Speech Quality and Efficient Vector Quantization in SELP, *Proceedings of IEEE International Conference on Acoustics, Speech, and Signal Processing*, vol. 1, pp. 155-158, New York, April, 1988.

[11] V. Ramamoorthy, N.S. Jayant, "Enhancement of ADPCM Speech by Adaptive Postfiltering," *Conf. Rec., IEEE Conf. on Commun.*, pp. 917-920, June 1985.

[12] Y. Yatsuzuka, S. Iizuka, T. Yamazaki, "A variable Rate Coding by APC with Maximum Likelihood Quantization from 4.8 bit/s to 16 kbit/s," *Proc. Inter. Conf. Acoust., Speech, & Signal Processing*, pp. 3071-3074, April 1986.

[13] J. H. Chen and A. Gersho, "Real-Time Vector APC Speech Coding at 4800 bps with Adaptive Postfiltering," *Proc. Int. Conf. on Acoust., Speech, Signal Processing Speech, and Signal Processing*, vol. 4, pp. 2185-2188, Dallas, April 1987.

[14] J.P. Campbell, Jr., V.C. Welch, T.E. Tremain, "An Expandable Error-Protected 4800 BPS CELP Coder (U.S. Federal Standard 4800 BPS Voice Coder)," *Proc. Inter. Conf. Acoust., Speech, & Signal Processing*, pp. 735-738, May 1989.

[15] V. Cuperman, A. Gersho, R. Pettigrew, J. Shynk, J. Yao and J. H. Chen, "Backward Adaptive Configurations for Low-Delay Speech Coding," *Proc., IEEE Global Commun. Conf.*, November 1989.

[16] J. H. Chen, "A Robust Low-Delay CELP Speech Coder at 16 kb/s," *Proc., IEEE Global Commun. Conf.*, November 1989.

[17] Shihua Wang and Allen Gersho, ''Phonetically-Based Vector Excitation Coding of Speech at 3.6 kbit/s,'' *Proc. IEEE Inter. Conf. Acoust., Speech, and Signal Processing*, Glasgow, May 1989.

[18] Shihua Wang and Allen Gersho, ''Phonetic Segmentation for Low Rate Speech Coding,'' *Advances in Speech Coding*, Kluwer Academic Publishers, to appear 1990..

[19] A. Gersho, ''Optimal Nonlinear Interpolative Vector Quantization,'' *IEEE Trans. on Comm.*, vol. COM-38, No. 9, pp. 1285-1287, September 1990.

ADAPTIVE PROCESSING IN SENSOR ARRAYS

M. Wax
Rafael, Haifa, Israel

ABSTRACT

We present the key concepts and techniques for detection, localization and beamforming of multiple narrowband sources by passive sensor arrays. We address the case of arbitrarily correlated sources, including the case of full correlation occuring in specular multipath propagation, and arbitrarily structured arrays.

1. INTRODUCTION

The topic of adaptive processing in sensor arrays arises in a variety of fields ranging from radar, sonar, oceanography and seismology to medical imaging and radio-astronomy, and have been the subject of numerous paprers and books (see Hudson (1981), Monzingo and Miller (1982), Haykin (1985), Pillai (1989)).

Sensor arrays are characterized by their spatial geometry and by the directional patterns of the sensors. If the geometry of the array and the directional patterns are very regular the array is referred to as structured, otherwise it is referred to as unstructured. Although structured arrays open the possiblity to sophisticated processing algorithms that exploit the structure, mutual coupling between the sensors and geometrical constraints preclude in many cases the possibility of implementing such arrays in practice.

The signals impinging on sensor arrays are chracterized by their bandwidth and by their mutual correlation. If the bandwidth of the signal is much smaller than the reciprocal of the propagation time across the array the signal is referred to as narrowband, otherwise it is referred to as wideband. Since wideband signals have in many cases narrowband components, the core of the research in sensor arrays have been centred on narrowband signals. The signals can be uncorrelated, as happens when the signals are radiated from two independent sources, or can be partially or fully correlated, as happens in multipath propagation.

The scenario we address is that of narrowband and arbitrarily correlated sources impinging on an arbitrarily structured array. For this rather general scenario we present optimal and suboptimal solutions to the problems of detection, localization and beamforming, which are to the key problems in adaptive processing in sensor arrays. In section 1 we present the mathematical and statistical models for the signals and and noise and formulate the problems. In Section 2 we presents the key concepts upon which the solutions are based. Then, in Section 3 we present the conditions that guarantee the uniqueness of the solutions. In section 4 and 5 we present the popular suboptimal non-model-based and model-based solutions. Then, in sections 6 and 7 we present the optimal solutions to the case of spatially white noise. Finally, in section 8 we present a suboptimal solution to the case of spatially colored noise.

1.1 THE MATHEMATICAL MODEL

Consider an array composed of p sensors with arbitrary locations and arbitrary directional characteristics. Assume that q narrowband sources, centered around a known frequency, say ω_0, impinge on the array from distinct locations $\theta_1, \ldots, \theta_q$. For simplicity, assume that the sources and the sensors are coplanar and that the sources are in the far-field of the array. In this case the only parameter that characterizes the source location is its direction-of-arrival θ.

Using complex envelope representation, the $p \times 1$ vector received by the array can be expressed by

$$\mathbf{x}(t) = \sum_{k=1}^{q} \mathbf{a}(\theta_k) s_k(t) + \mathbf{n}(t), \tag{1.1}$$

where $\mathbf{a}(\theta)$ is the $p \times 1$ "steering vector" of the array towards direction θ,

$$\mathbf{a}(\theta) = \left[a_1(\theta) e^{-j\omega_0 \tau_1(\theta)}, \ldots, a_p(\theta) e^{-j\omega_0 \tau_p(\theta)} \right]^T, \tag{1.2}$$

$a_k(\theta)-$ denotes the amplitude response of the k-th sensor towards direction θ,

$\tau_k(\theta)-$ denotes the propagation delay between the reference point and the k-th sensor to a wavefront impinging from direction θ,

$s_k(t)$– denotes the signal of the k-th source as received at the reference point,

and $\mathbf{n}(t)$ denotes the $p \times 1$ vector of the noise at the sensors.

In matrix notation this becomes

$$\mathbf{x}(t) = \mathbf{A}(\Theta)\mathbf{s}(t) + \mathbf{n}(t), \tag{1.3}$$

where $\mathbf{A}(\Theta)$ is the $p \times q$ matrix of the steering vectors,

$$\mathbf{A}(\Theta) = \left[\mathbf{a}(\theta_1), \ldots, \mathbf{a}(\theta_q)\right]. \tag{1.4}$$

Suppose that the received vector $\mathbf{x}(t)$ is sampled M times, at $t_1, \ldots, t_M$. From (1.1), the sampled data can be expressed as

$$\mathbf{X} = \mathbf{A}(\Theta)\mathbf{S} + \mathbf{N}, \tag{1.5}$$

where $\mathbf{X}$ and $\mathbf{N}$ are the $p \times M$ matrices

$$\mathbf{X} = [\mathbf{x}(t_1), \ldots, \mathbf{x}(t_M)], \tag{1.6}$$

$$\mathbf{N} = [\mathbf{n}(t_1), \ldots, \mathbf{n}(t_M)], \tag{1.7}$$

and $\mathbf{S}$ is the $q \times M$ signals matrix

$$\mathbf{S} = [\mathbf{s}(t_1), \ldots, \mathbf{s}(t_M)]. \tag{1.8}$$

1.2 THE STATISTICAL MODELS

There are two common statistical models for the noise. One is referred to as the White Noise (WN) model and the other as the Colored Noise (CN) model.

WN: The noise samples $\{\mathbf{n}(t_i)\}$ are statistically independent Gaussian random vectors with zero mean and covariance matrix $\sigma^2\mathbf{I}$.

CN: The noise samples $\{\mathbf{n}(t_i)\}$ are statistically independent Gaussain random vectors with zero mean and covariance matrix $\mathbf{Q}$.

Similarly, there are two common statistical models for the signals. One is referred to as the Deterministic Signals (DS) model and the other as the Stochastic Signals (SS) model.

DS: the signals $\{\mathbf{s}(t_i)\}$ are regarded as unknown deterministic constants.

SS: The signal samples $\{\mathbf{s}(t_i)\}$ are independent and identical Gaussain random vectors with zero mean and covariance matrix $\mathbf{P}$.

In both models, the shape of the signals and the correlation amongst them can be arbitrary. Specifically, the signals can be fully correlated as happens, for example, in specular multipath propagation.

1.3 PROBLEM FORMULATION

The problem can be stated as follows. Given the sampled data $\mathbf{X}$, determine:

(i) the number of sources q.

(ii) the directions-of-arrival $\theta_1, \ldots, \theta_q$.

(iii) the signal waveforms $\mathbf{s}(t_1), \ldots \mathbf{s}(t_M)$.

We shall refer to (i) as the detection problem, to (ii) as the localization problem and to (iii) as the beamforming problem.

1.4 MATHEMATICALLY EQUIVALENT PROBLEMS

The mathematical model of the sensor array problem is identical to the model which arises in other key problems in signal processing. The solutions we shall describe to the sensor array problem are therefore applicable to these problems as well.

1.4.1 Harmonic Retrieval

Consider a signal $x(t)$ composed of q sinusoids with unknown parameters embedded in additive noise:

$$x(t) = \sum_{k=1}^{q} a_k \cos(\omega_k t + \phi_k) + n(t), \tag{1.9}$$

where ω_k, a_k and ϕ_k are the frequency, amplitude and phase, respectively, of the k-th sinusoid.

Let a tapped-delay-line with p equally spaced taps D delay units apart be used to sample the signal. The signal at the $(i+1)$-th tap is given by

$$x(t - iD) = \sum_{k=1}^{q} a_k \cos(\omega_k (t - iD)) + \phi_k) + n_i(t), \tag{1.10}$$

Using complex representation we can express the $p \times 1$ vector of the taped-delay-line outputs as

$$\mathbf{x}(t) = \sum_{k=1}^{q} \mathbf{a}(\omega_k) s_k(t) + \mathbf{n}(t), \tag{1.11}$$

where $s_k(t)$ is the complex sinusoid

$$s_k(t) = a_k e^{j(\omega_k t + \phi_k)}, \tag{1.12}$$

and $\mathbf{a}(\omega_k)$ is the $p \times 1$ vector

$$\mathbf{a}(\omega_k) = [1, e^{-j\omega_k D}, \ldots, e^{-j\omega_k (p-1)D}]^T, \tag{1.13}$$

which is exactly in the form of (1.1).

1.4.2 Pole Retrieval

Consider a linear system that is excited by an impulse. Assuming that the system has q unknown and distinct poles in the complex plane, the response of the system can be expressed as

$$x(t) = \sum_{k=1}^{q} a_k e^{\alpha_k} \cos(\omega_k t + \phi_k) + n(t), \tag{1.14}$$

where $s_k = \alpha_k + j\omega_k$ is the location of the k-th pole in the complex plane and a_k and ϕ_k are its residue and phase, respectively.

Let a tapped-delay-line with p equally spaced taps D delay units apart be used to sample the signal. Using complex representation we can express the $p \times 1$ vector of the taped-delay-line outputs as

$$\mathbf{x}(t) = \sum_{k=1}^{q} \mathbf{a}(s_k) h_k(t) + \mathbf{n}(t), \tag{1.15}$$

where $h_k(t)$ is the response to the k-th pole

$$h_k(t) = a_k e^{s_k t + \phi_k}, \tag{1.16}$$

and $\mathbf{a}(s_k)$ is the $p \times 1$ vector

$$\mathbf{a}(s_k) = [1, e^{-s_k D}, \ldots, e^{-s_k (p-1) D}]^T, \tag{1.17}$$

which is exactly in the form of (1.1).

1.4.3 Echo Retrieval

Consider a radar or sonar system that transmits a *known* pulse $s(t)$ and recieves a backscattered signal. Assume that the backscattered signal can be modeled as a superposition of q scaled and delayed echoes embedded in additive noise:

$$x(t) = \sum_{k=1}^{q} m_k s(t - \tau_k) + n(t), \tag{1.18}$$

where m_k and τ_k are the amplitude and the delay of the k-th echo, respectively.

Let a tapped-delay-line with p equally spaced taps D delay units apart be used to sample the signal. The $p \times 1$ vector of the taped-delay-line outputs can be expressed as

$$\mathbf{x}(t) = \sum_{k=1}^{q} \mathbf{a}(\tau_k) m_k(t) + \mathbf{n}(t), \tag{1.19}$$

where $\mathbf{a}(\tau_k)$ is the $p \times 1$ vector

$$\mathbf{a}(\tau_k) = [s(t - \tau_k), \ldots, s(t - (p-1)D - \tau_k)]^T, \tag{1.20}$$

which is exactly in the form of (1.1).

2. PRELIMINARIES

In this section we present the fundamental concepts upon which the solutions will be based.

2.1 THE SIGNAL SUBSPACE

To gain some insight into the nature of the problems at hand, we cast them in a geometrical framework first presented by Schmidt (1979, 1981).

Observe that the p-dimensional complex vector $\mathbf{x}(t)$ can be regarded as a point in $\mathbf{C}^p$. The location of $\mathbf{x}(t)$ in $\mathbf{C}^p$ is determined by the constraints imposed by (1.1). To see the nature of these constraints we consider two simple cases.

Consider first the case of a single source with no noise present. In this case (1.1) becomes

$$\mathbf{x}(t) = \mathbf{a}(\theta_1)s_1(t), \tag{2.1}$$

implying that $\mathbf{x}(t)$ is confined to a 1-dimensional subspace spanned by $\mathbf{a}(\theta_1)$.

Similarly, for the case of two sources with no noise present (1.1) becomes

$$\mathbf{x}(t) = \mathbf{a}(\theta_1)s_1(t) + \mathbf{a}(\theta_2)s_2(t), \tag{2.2}$$

implying that $\mathbf{x}(t)$ is confined to 2-dimensional subspace spanned by $\mathbf{a}(\theta_1)$ and $\mathbf{a}(\theta_2)$. The path traced by $\mathbf{x}(t)$ in this plane depends on the values of $s_1(t)$ and $s_2(t)$ and on their statistical correlation. If $s_1(t)$ and $s_2(t)$ are uncorrelated then $\mathbf{x}(t)$ traces a random pattern in the plane. If, on the other hand, $s_1(t)$ and $s_2(t)$ are fully correlated, i.e., $s_1(t) = cs_2(t)$, as happens in specular multipath propagation, the path traced by $\mathbf{x}(t)$ is confined to the line $\mathbf{a}(\theta_1) + c\mathbf{a}(\theta_2)$ in this plane.

In general, in the case of q sources with no noise present (1.1) becomes

$$\mathbf{x}(t) = \sum_{i=1}^{q} \mathbf{a}(\theta_i)s_i(t), \tag{2.3}$$

i.e, $\mathbf{x}(t)$ is confined to the q-dimensional subspace span$\{\mathbf{a}(\theta_1),\ldots,\mathbf{a}(\theta_q)\}$. From obvious reasons, this subspace is referred to as the "signal subspace," and its complement is referred to as the "noise subspace." If the signals are uncorrelated then $\mathbf{x}(t)$ spans the whole signal subspace while if part of the signals or all of them are fully correlated then $\mathbf{x}(t)$ spans only part of the signal subspace.

The solution of the localization problem in the absence of noise amounts therefore to finding the *smallest* number q and the locations $\theta_1,\ldots,\theta_q$ for which the signal subspace, span$\{\mathbf{a}(\theta_1),\ldots,\mathbf{a}(\theta_q)\}$, *contains* the sampled data $\{\mathbf{x}(t_i)\}$.

In the presence of noise, the vectors $\mathbf{x}(t)$ are no longer confined to the signal subspace and instead are scattered around this subspace. The solution of the localization problem under these more realistic conditions is therefore much more subtle. Intuitively, it amounts to finding the *smallest* number q and the locations $\theta_1,\ldots,\theta_q$ for which span$\{\mathbf{a}(\theta_1),\ldots,\mathbf{a}(\theta_q)\}$ " best fits" the sampled data $\{\mathbf{x}(t_i)\}$.

2.2 THE MDL MODEL SELECTION CRITERION

The measure for the "goodness-of-fit" between the potential signal subspace and the observed data will be derived from the information theoretic criteria for model selection pioneered by Akaike (1973, 1974) and further developed by Schwartz (1978) and Rissanen (1978, 1983, 1989). These criteria, inspite of their very similar nature and form, are derived from quite different principles. Since we believe that Rissanen's approach is the most powerful operationally and conceptually, we base our derivations on his Minimum Description Length (MDL) principle.

The MDL principle asserts: given a data set and a family of competing statistical models, the best model is the one that yields the ***minimal description length*** of the data. The rationale is that since a good model is judged by its ability to "summarize" the data, the shorter is the "summary," the better is the corresponding model.

The description length of the data, i.e., the number of bits it takes to encode the data, can be evaluated quantitatively. Indeed, suppose we want to encode the data set $\mathbf{X} = \{\mathbf{x}(t_1), \ldots, \mathbf{x}(t_M)\}$ using the probabilistic model $P(\mathbf{X}|\mu)$, where μ is an *unknown* (vector) parameter. One possible encoding scheme, referred to as "two part coding," is to first encode an estimate of the parameter vector μ and then encode the data set $\mathbf{X}$ using the value of the encoded parameter. The code length in this scheme is therefore given by

$$L(\mathbf{X}) = L(\tilde{\mu}) + L(\mathbf{X}|\tilde{\mu}), \tag{2.4}$$

where $L(\tilde{\mu})$ is the code length of the parameter vector and $L(\mathbf{X}|\tilde{\mu})$ is the code length of the data.

To encode the parameters we use a simple coding scheme. Suppose the value of the estimated parameter vector is $\hat{\mu}$ and suppose we encode the parameters by truncating them to some precision, say $\hat{\mu}_i$ to the precision $\delta_i = 2^{-q_i}$, where q_i is the number of fractional binary digits taken in the truncation. In this case

$$L(\tilde{\mu}) = \sum_{i=1}^{\nu} - \log \delta_i, \tag{2.5}$$

where ν denotes the number of free parameters in μ.

It was shown already by Shannon (1948) that if the parameter μ is known then the optimal code length for the data is given by $-\log P(\mathbf{X}|\mu)$. Since the value of μ is unknown, it is natural to look for the value of μ for which the code length is minimized. This yields

$$\min_{\mu}\{-\log P(\mathbf{X}|\mu)\} = \max_{\mu} \log P(\mathbf{X}|\mu), \tag{2.6}$$

which is the maximum likelihood criterion, i.e.,

$$L(\mathbf{X}|\tilde{\mu}) = -\log P(\mathbf{X}|\tilde{\mu}), \tag{2.7}$$

where $\tilde{\mu}$ is the truncated value of the maximum likelihood estimator $\hat{\mu}$.

Combining the two code lengths, we get

$$L(\mathbf{X}) = -\log P(\mathbf{X}|\tilde{\mu}) + \sum_{i=1}^{\nu} - \log \delta_i. \tag{2.8}$$

Not surprisingly, this code length depends on the precision selected for the parameters. While the last term increases as the precision increases since the $\{\delta_i\}$ are smaller, the first term decreases since the truncated parameters better approximate the optimal value $\hat{\mu}$. Consequently, there is a certian precision for which the code length is minimized.

To find this optimal precision, we expand the first term in a Taylor series around the optimal value $\hat{\mu}$. Since the first derivative vanishes at the optimal value we get

$$L(\mathbf{X}|\delta) = -\log P(\mathbf{X}|\hat{\mu}) + \frac{1}{2}\delta^T \Sigma \delta + \sum_{i=1}^{\nu} -\log \delta_i. \tag{2.9}$$

where $\delta = [\delta_1, \ldots, \delta_\nu]^T$ and Σ is the Hessian matrix of $-\log P(\mathbf{X}|\hat{\mu})$.

Differentiation of the right hand side with respect to δ we get that the minimum is achieved for the value of δ given by the equation

$$\Sigma \delta = \delta^{(-1)}, \tag{2.10}$$

where

$$\delta^{(-1)} = [\frac{1}{\delta_1}, \ldots, \frac{1}{\delta_\nu}]^T. \tag{2.11}$$

Substituting the optimizing precision in the expression for the code length we get

$$L(\mathbf{X}|\hat{\delta}) = -\log P(\mathbf{X}|\hat{\mu}) + \frac{1}{2}\nu + \sum_{i=1}^{\nu} -\log \hat{\delta}_i. \tag{2.12}$$

To further simplify this expression we consider its asymptotic behaviour. Assuming that $-\log P(\mathbf{X}|\mu)$ grows proportionally to the number of samples M, as is normally the case, the elements of $\frac{1}{M}\Sigma$ are of order 1, regardless of M. Equation (2.10) then implies that

$$\hat{\delta}_i = \frac{c_i(M)}{\sqrt(M)}, \tag{2.13}$$

where $\{c_i(M)\}$ are some bounded parameters, and hence the description length of the data is asymptotically given by

$$L(\mathbf{X}) = -\log P(\mathbf{X}|\hat{\mu}) + \frac{1}{2}\nu \log M + O(\nu), \tag{2.14}$$

where

$$O(\nu) = \frac{\nu}{2} + \sum_{i=1}^{\nu} \log c_i(M). \tag{2.15}$$

Assuming M is large and $\frac{\nu}{\log M} \to 0$, the Minimum Description Length (MDL) is given by

$$MDL = -\log P(\mathbf{X}|\hat{\mu}) + \frac{1}{2}\nu \log M, \tag{2.16}$$

Notice that since the asymptotic analysis is valid only if $\frac{\nu}{\log M} \to 0$, the MDL model selection criterion (2.16) is not applicable to the pathological case wherein the number of free parameters grows linearly with the number of samples.

It is appropriate at this point to compare the MDL model selection criterion with the AIC. This criterion, based on an asymptotic analysis of the Kulback-Liebler distance between the "true" and estimated models, is given by

$$AIC = -\log P(\mathbf{X}|\hat{\mu}) + \nu. \tag{2.17}$$

Notice that the only difference between the two criteria is in the coefficient multiplying the number of free parameters in the second term; it is $\frac{1}{2}\log M$ in the MDL and 1 for the AIC.

It turns out that this seemingly minor difference has important implications in terms of the asymtotic behaviour of the two criteria. Indeed, in all the analyzed models the MDL was proved to be consistent, i.e., it selects the true model with probability one when the number of samples grows to infinity, while the AIC was proved to be inconsistent, with a tendency to select more complex models then the true one.

It should be pointed out that the consistency property is shared by a larger family of Information Criteria. Indeed any Information Criterion with the following generic structure

$$IC = -\log P(\mathbf{X}|\hat{\mu}) + \alpha(M)\log M. \tag{2.18}$$

can be shown to be consistent, Zhao *et. al.* (1986), if $\alpha(M) \to \infty$ and $\frac{\alpha(M)}{M} \to 0$ as $M \to \infty$.

3. CONDITIONS FOR UNIQUENESS

In this section we consider the uniqueness of the solution. Our presentation follows Wax and Ziskind (1989a) with a slight simplification due to Nehorai, Starer and Stoica (1989).

By its nature, the uniqueness problem is decoupled from the estimation problem. Thus, we can ignore the noise in the following analysis and rewrite (1.5) as

$$\mathbf{X} = \mathbf{A}(\Theta)\mathbf{S}. \tag{3.1}$$

Expression (3.1) can be regarded as a set of equations in which the left hand side contains the given data while the right hand side contains the unknowns. Our goal is to specify the conditions under which the solution $(\Theta, \mathbf{S})$ of this set of equations is unique.

Referring to the set $\{\mathbf{a}(\theta), \theta \in \Omega\}$, where Ω denotes the field-of-view, as the " array manifold," the conditions we impose on the array are given by:

A1: The array manifold is known.

A2: Any subset of p distinct steering vectors from the array manifold is linearly independent.

A1 can be fulfilled by either computing the array manifold analytically, if possible, or, alternatively, by measuring it *in situ*. A2 imposes mild constraints on the array geometry which can easily be fulfilled.

To specify the conditions on the sources, let η denote the rank of the $q \times M$ signal matrix $\mathbf{S}$,

$$\eta = \text{rank } \mathbf{S}. \tag{3.2}$$

Note that

$$\eta = \text{rank } [\mathbf{S}\mathbf{S}^H] = \text{rank}[\sum_{i=1}^{M} \mathbf{s}(t_i)\mathbf{s}(t_i)^H], \tag{3.3}$$

where H denotes the conjugate transpose, which implies that η is the rank of the sample covariance of the signals,

$$\eta = \text{rank}[\frac{1}{M}\sum_{i=1}^{M} \mathbf{s}(t_i)\mathbf{s}(t_i)^H]. \tag{3.4}$$

To point out the implication of $\mathbf{S}$ being of rank η, we rewrite (3.1) as

$$[\mathbf{X}_1, \mathbf{X}_2] = \mathbf{A}(\Theta)[\mathbf{S}_1, \mathbf{S}_2], \tag{3.5}$$

where $\mathbf{X}_1$ and $\mathbf{S}_1$ are $p \times \eta$ matrices, and $\mathbf{X}_2$ and $\mathbf{S}_2$ are $p \times (M - \eta)$ matrices. Now, since the matrix $\mathbf{S}$ is of rank η we can assume, without loss of generality, that $\mathbf{S}_1$ is full column rank and that

$$\mathbf{S}_2 = \mathbf{S}_1\mathbf{B}, \tag{3.6}$$

where $\mathbf{B}$ is the matrix expressing the linear dependence. From (3.5) and (3.6) we get

$$\mathbf{X}_2 = \mathbf{A}(\Theta)\mathbf{S}_1\mathbf{B} = \mathbf{X}_1\mathbf{B}, \tag{3.7}$$

and hence the system of equations (3.5) is equivalent to the reduced system

$$\mathbf{X}_1 = \mathbf{A}(\Theta)\mathbf{S}_1, \tag{3.8}$$

so that we can confine the analysis to $p \times \eta$ matrices $\mathbf{S}$ of full column rank.

We shall consider two types of uniqueness: (i) uniqueness for every batch $\mathbf{X}$, (ii) uniqueness for almost every batch $\mathbf{X}$, i.e., with the exception of a set of batches $\mathbf{X}$ of measure zero.

We first state the conditions that guarantee uniqueness for every batch $\mathbf{X}$.

Theorem 1:

An array satisfying conditions A1 and A2 can always, that is for every batch $\mathbf{X}$, uniquely localize q sources provided that

$$q < \frac{p+\eta}{2}. \tag{3.9}$$

Proof:

We shall show that if (3.9) holds true then for every $\mathbf{X}$

$$\mathbf{X} = \mathbf{A}(\Theta)\mathbf{S} \neq \mathbf{A}(\Theta')\mathbf{S}', \tag{3.10}$$

for any locations $\Theta' \neq \Theta$ and any set of signals $\mathbf{S}'$.

To this end, suppose $\theta_i = \theta_j'$ for d pairs of i and j $(i, j = 1, ..., q)$. Observe that the case $\theta_i \neq \theta_j'$ for all i and j $(i, j = 1, ..., q)$ corresponds to $d = 0$.

In this case we can rewrite (3.10) as

$$[\mathbf{A}(\Theta), \tilde{\mathbf{A}}(\Theta')] \begin{bmatrix} \tilde{\mathbf{S}} \\ -\tilde{\mathbf{S}}' \end{bmatrix} \neq 0, \tag{3.11}$$

where $\tilde{\mathbf{A}}(\Theta')$ denotes the $p \times (q-d)$ matrix obtained from $\mathbf{A}(\Theta')$ by deleting those d columns which are identical to the columns of $\mathbf{A}(\Theta)$, $\tilde{\mathbf{S}}'$ denotes the

$(q-d) \times \eta$ matrix obtained from $\mathbf{S}'$ by deleting the d rows corresponding to the deleted columns in $\mathbf{A}(\Theta')$, and $\tilde{\mathbf{S}}$ denotes the $q \times \eta$ matrix obtained from $\mathbf{S}$ by subtracting the deleted rows of $\mathbf{S}'$ from the corresponding ones in $\mathbf{S}$. Denote

$$\varsigma = \text{null}[\mathbf{A}(\Theta), \tilde{\mathbf{A}}(\Theta')] = (2q-d) - \text{rank}[\mathbf{A}(\Theta), \tilde{\mathbf{A}}(\Theta')], \tag{3.12}$$

and

$$\nu = \text{rank}\begin{bmatrix} \tilde{\mathbf{S}} \\ -\tilde{\mathbf{S}}' \end{bmatrix}. \tag{3.13}$$

It follows from (3.11) that to prove the theorem it suffices to show that

$$\varsigma < \nu. \tag{3.14}$$

To this end, first note that since rank$\mathbf{S} = \eta$ and since $\tilde{\mathbf{S}}$ is obtained from $\mathbf{S}$ by deleting d rows then rank$\tilde{\mathbf{S}} \geq \eta - d$ and hence we have

$$\text{rank}\begin{bmatrix} \tilde{\mathbf{S}} \\ -\tilde{\mathbf{S}}' \end{bmatrix} \geq \eta - d \geq 1. \tag{3.15}$$

Next, by A2 we have

$$\text{rank}[\mathbf{A}(\Theta), \tilde{\mathbf{A}}(\Theta')] = \min\{p, 2q-d\}. \tag{3.16}$$

Now if $d < 2q - p$ then (3.16) implies that $\varsigma = 2q - p - d$ and hence since from (3.9) $2q - p - d < \eta - d$ and from (3.15) $\nu \geq \eta - d$ it follows that $\varsigma < \nu$.

Similarly, if $d \geq 2q - p$ then it follows from (3.16) that $\varsigma = 0$ and since from (3.15) $\nu \geq 1$ we get that $\varsigma < \nu$. This completes the proof.

It is of interest to examine condition (3.9) for some special cases. One important case is $\eta = q$, occuring when the sources are uncorrelated. In this case (3.9) implies that a unique solution is guaranteed for every batch of data if $q < p$. Another important case is $\eta = 1$, occuring when the sources are fully correlated (as happens in specular multipath propagation), and in the case of a single snapshot, i.e., when M=1. In this case (3.9) implies that a unique solution is guaranteed for every batch of data if $q < (p+1)/2$.

Theorem 1 states the conditions required to guarantee uniqueness for *every* batch **X**. However, condition (3.9) is rather demanding and cannot be fulfilled in many applications. It is of interest therefore whether this condition can be weakend so as to guarantee uniqueness for *almost every* batch **X**, i.e., with the exception of a set of batches of measure zero.

To formalize this approach we introduce the notion of "topological dimension" of a set. This notion is rigorously defined in Hurewicz and Whallman (1948). However in order not to clutter the discussion with inessential technical details, we adopt the following definition which is sufficient for our problem. A set S is said to be m-dimensional if m is the least number of real parameters needed to describe its points.

Lemma 1:

Let $S(\Theta)$ be a m-dimensional set, with Θ denoting a $n \times 1$ parameter vector in R^n, and let S denote the union of all possible $S(\Theta)$'s,

$$S = \cup_{\Theta} S(\Theta). \tag{3.17}$$

Then

$$\dim S \leq m + n. \tag{3.18}$$

Proof:

In order to prove the lemma it suffices to show that the set S can be described with no more than $m+n$ real parameters. Indeed, for each point P in the set B, n real parameters are needed to specify Θ and hence the subset $S(\Theta)$ in which P lies, and m real parameters are needed to specify the coordinates of P inside $S(\Theta)$. This amounts to a total of $m+n$ real parameters and hence the dimension of S is less than or equal to $m+n$.

Having established the above result from dimension theory we now state the conditions that guarantee the uniqueness for almost every batch $\mathbf{X}$.

Theorem 2:

Let Θ be fixed and let $\mathbf{S}$ be a $q \times \eta$ random matrix, drawn from the set of all rank-η matrices whose elements are jointly distributed according to some absolutely continuous distribution on $R^{2q\eta}$. An array satisfying A1 and A2 can then, with probability one, uniquely localize q sources provided that

$$q < \frac{2\eta}{2\eta+1}p. \tag{3.19}$$

Proof:

Let the set of all matrices $\mathbf{S}$ for which Θ and Θ' are two ambiguous solutions, where Θ' is arbitrary, be denoted by $S(\Theta,\Theta')$,

$$S(\Theta,\Theta') = \{\mathbf{S},\ \mathbf{A}(\Theta)\mathbf{S} = \mathbf{A}(\Theta')\mathbf{S}'\}, \tag{3.20}$$

Here $\mathbf{S}'$ denotes some $q \times \eta$ matrix. Finally, let $S(\Theta)$ denote the union of the sets $S(\Theta,\Theta')$, i.e., the set of all matrices $\mathbf{S}$ for which ambiguous solutions exist,

$$S(\Theta) = \cup_{\Theta'} S(\Theta,\Theta'). \tag{3.21}$$

Our proof is based on comparing the topological dimension of the set S to that of the set $S(\Theta)$. To this end, observe that $q\eta$ complex parameters are required to describe a (rank-η) matrix $\mathbf{S}$ in S. Therefore

$$\dim S = 2q\eta. \tag{3.22}$$

To derive the topological dimension of the set $S(\Theta)$, we have to evaluate the topological dimension of the set $S(\Theta,\Theta')$. To this end, as in the proof of Theorem 1, suppose that $\theta_i = \theta_j'$ for d pairs of i and j $(i,j = 1,\dots,q)$.

In this case $\text{null}[\mathbf{A}(\Theta),\mathbf{A}(\Theta')]$ is given by (3.12). This implies that $\varsigma\eta$ complex parameters are required to describe a (rank-η) matrix $\mathbf{S}$ in $S(\Theta,\Theta')$ and therefore

$$\dim S(\Theta,\Theta') = 2\varsigma\eta. \tag{3.23}$$

Now, if $d < 2q - p$ we get from (3.16) that

$$\dim S(\Theta,\Theta') = 2(2q-p)\eta. \tag{3.24}$$

Using Lemma 1, we then have

$$\dim S(\Theta) \leq 2(2q-p)\eta + q. \tag{3.25}$$

Now if (3.19) holds true then $2(2q-p)\eta + q < 2q\eta$. This then implies, by (3.25) and (3.22), that

$$\dim S(\Theta) < \dim S. \tag{3.26}$$

Similarly, if $d \geq 2q - p$ it follows from (3.16) that

$$\dim S(\Theta,\Theta') = 2d\eta. \tag{3.27}$$

Using Lemma 1 this yields

$$\dim S(\Theta) \leq 2d\eta + (q-d). \tag{3.28}$$

Now evidently $2(q-d)\eta > (q-d)$ and hence $2d\eta + q - d < 2q\eta$. This then implies, by (3.28) and (3.22), that

$$\dim S(\Theta) < \dim S. \tag{3.29}$$

We have thus shown that the topological dimension of the set $S(\Theta)$ is smaller than that of S. That is, $S(\Theta)$ is a *proper lower dimensional subset* of S. Now since our stochastic setting defines an absolutely continuous measure on the set S, the total measure induced on a proper lower dimensional subset of S is zero. Consequently, the probability of an ambiguous solution is zero. This completes the proof.

Observe that Theorem 2 does not exclude the possibility of ambiguous solutions. In fact, if $(p+\eta)/2 \leq q < (2\eta p)/(2\eta+1)$ then for every location vector Θ there exist some matrices $\mathbf{S}$, and hence some batches $\mathbf{X}$, for which the solution is ambiguous. What Theorem 2 assures is that the measure of this set of batches within the set of all possible batches, when $\mathbf{S}$ is drawn at random from the set of rank-η $q \times \eta$ matrices, is zero.

Observe that in the special case of coherent sources, $\eta = 1$, condition (3.19) becomes $q < 2p/3$. It should be pointed out that Bresler and Macovski (1986) proved that for the case of a uniform linear array this condition is $q \leq 2p/3$, which is slightly less demanding.

Theorems 1 and 2 specify sufficient conditions for uniqueness. It turns out, that a slightly weaker version of (3.19) can be shown to be necessary. Indeed, we now show that a necessary condition for uniqueness for every batch $\mathbf{X}$ is

$$q \leq \frac{2\eta}{2\eta+1} p. \tag{3.30}$$

Suppose that (3.30) does not hold, namely

$$2p\eta < 2q\eta + q. \tag{3.31}$$

The right hand side of (3.31) represents the number of unknowns, while the left hand side represents the number of independent equations. Thus, (3.31) implies that equation (3.2) has more unknowns than equations and hence, by the implicit function theorem (see e.g. Rudin (1976)), assuming that the conditions for its existence hold, an infinite number of ambiguous solutions exist in the proximity of the solution $(\Theta, \mathbf{S})$. This establishes the necessity of (3.30) for uniqueness.

In Wax and Ziskind (1989a) it is conjectured that the necessary condition (3.30) is also sufficient to establish uniqueness with probability one. In fact, for the special case of a uniform linear array and $\eta = 1$, the sufficiency of (3.30) has been established already by Bresler and Macovski (1986).

4. SUBOPTIMAL NON-MODEL-BASED SOLUTIONS

In this section we present the "classical" solutions which do not exploit the detailed nature of the model of $\mathbf{x}(t)$ nor the statistical models of the signals and noise.

4.1 DELAY-AND-SUM

The oldest and still one of the most common solutions in use today is the "delay-and-sum" technique, referred to also as the classical beamformer. To motivate this technique suppose that a wavefront impinges on the array from direction θ and we want to coherently sum the received signals at the different sensors. To this end we have to steer a beam towards direction θ, namely first properly delay the received signals at the different sensors and then sum them. This is carried out by

$$\mathbf{w}(\theta)^H \mathbf{x}(t), \tag{4.1}$$

where $\mathbf{w}(\theta)$ denotes the vector of delays (phase shifts) causing coherent summation from direction θ,

$$\mathbf{w}(\theta) = [e^{-j\omega_0\tau_1(\theta)}, \ldots, e^{-j\omega_0\tau_p(\theta)}]^T. \tag{4.2}$$

The average power at the delay-and-sum processor is given by

$$B(\theta) = \frac{1}{M}\sum_{i=1}^{M} \| \mathbf{w}(\theta)^H \mathbf{x}(t_i) \|^2, \tag{4.3}$$

which can be rewritten as

$$B(\theta) = \mathbf{w}(\theta)^H \hat{\mathbf{R}} \mathbf{w}(\theta), \tag{4.4}$$

where $\hat{\mathbf{R}}$ denots the sample covariance matrix

$$\hat{\mathbf{R}} = \frac{1}{M}\sum_{i=1}^{M} \mathbf{x}(t_i)\mathbf{x}(t_i)^H. \tag{4.5}$$

In order to solve the stated problem we have therefore to first compute $B(\Theta)$ for every θ, then determine the locations of the sources as the $\hat{q}$ peaks of $B(\theta)$ that pass a certain threshold, and finally steer beams to these $\hat{q}$ directions to estimate the signal waveforms.

Evidently, the delay-and-sum technique is suited for a *single* source; in this case $B(\theta)$ will have, asymptotically, a single maximum at the correct direction. However, if more then one source is present, the peaks may be at the wrong directions or may not be resolved at all because of the relatively poor resolution properties of this technique.

4.2 MINIMUM VARIANCE

The minimum variance technique was proposed by Capon (1969) to overcome the resolution problems of the delay-and-sum solution. Capon realized that the poor resolution of the delay-and-sum technique can be attributed to the fact that the power of the delay-and-sum processor at a given direction does not depend only on the power of the source at that direction, but also on undesireable contributions from other sources. To improve the resolution, he proposed to modify the delay-and-sum technique so as to minimize these interferences.

His approach was as follows. Suppose we want to receive a wavefront from direction θ and carry it out by a linear combination of the received signals, that is by

$$\mathbf{w}(\theta)^H \mathbf{x}(t), \tag{4.6}$$

where $\mathbf{w}(\theta)$ is a $p \times 1$ complex vector to be determined. It follows from (1.1) that to sum the wavefront from direction θ, the vector $\mathbf{w}$ must obey

$$\mathbf{w}(\theta)^H \mathbf{a}(\theta) = 1. \tag{4.7}$$

The output power of the linear combiner is given by

$$C(\theta) = \frac{1}{M} \sum_{i=1}^{M} \| \mathbf{w}(\theta)^H \mathbf{x}(t_i) \|^2, \tag{4.8}$$

which can be rewritten as

$$C(\theta) = \mathbf{w}(\theta)^H \hat{\mathbf{R}} \mathbf{w}(\theta), \tag{4.9}$$

where $\hat{\mathbf{R}}$ denots the sample covariance matrix.

Now in order to minimize the contribution to the output power of other sources at directions different from θ, Capon proposed to select the vector $\mathbf{w}(\theta)$

so as to minimize the output power, i.e.,

$$\min_{\mathbf{w}(\theta)^H \mathbf{a}(\theta)=1} \mathbf{w}(\theta)^H \hat{\mathbf{R}} \mathbf{w}(\theta). \tag{4.10}$$

The solution of this minimization problem, obtained easily by the Lagrange-multiplier technique, is given by

$$\hat{\mathbf{w}}(\theta) = \alpha \hat{\mathbf{R}}^{-1} \mathbf{a}(\theta), \tag{4.11}$$

where α is a positive scalar given by

$$\alpha = \frac{1}{\mathbf{a}(\theta)^H \hat{\mathbf{R}}^{-1} \mathbf{a}(\theta)}. \tag{4.12}$$

The minimization of the power output, dictated by (4.10), may create serious difficulties when the interfering signals are *correlated* with the desired signals. In this case, the resulting vector $\hat{\mathbf{w}}$ will operate on the interfering signals so as to partially or fully cancel the desired signals, as dictated by the minimization requirement. Note that constraint on the vector $\mathbf{w}$ does not prevent this phenomenon since although the desired signal is summed coherently, the interfering signals are phase-shifted and scaled by $\mathbf{w}$ so as to partially or fully cancel the desired signal. This phenomenon is referred to as signal cancellation by Widrow *et al* (1982).

The average power at the output of the linear combiner obtained by using the optimal vector $\hat{\mathbf{w}}$ is

$$C(\theta) = \frac{1}{\mathbf{a}(\theta)^H \hat{\mathbf{R}}^{-1} \mathbf{a}(\theta)}. \tag{4.13}$$

Thus to solve our problems we have first to compute $C(\theta)$ for every θ, then determine the locations of the sources as the $\hat{q}$ peaks of $C(\theta)$ that pass a certain threshold, and finally steer beams to these $\hat{q}$ directions to estimate the signal waveforms.

4.2.1 Maximum Signal-to-Interference Interpretation

We shall now show that in the case the interefering signals are uncorrelated with the desired signals the Minimum Variance technique can be interpreted also as maximizing the signal-to-interference ratio at the beamformer output.

Indeed, let us first rewrite (1.1) as

$$\mathbf{x}(t) = \mathbf{a}(\theta)s(t) + \imath(t), \tag{4.14}$$

where $\imath(t)$ represents the interference due to the other impinging signals and the thermal noise.

The average of the power of the desired signal at the beamformer output is given by

$$E \parallel \mathbf{w}(\theta)^H \mathbf{a}(\theta)s(t) \parallel^2 = p\mathbf{w}(\theta)^H \mathbf{a}(\theta)\mathbf{a}(\theta)^H \mathbf{w}(\theta), \tag{4.15}$$

where p denotes the average power of the signal $s(t)$. Similarly, the average power output of the interference is given by

$$E \parallel \mathbf{w}(\theta)^H \imath(t) \parallel^2 = \mathbf{w}(\theta)^H \mathbf{R}_I \mathbf{w}(\theta), \tag{4.16}$$

where $\mathbf{R}_I$ denotes the average power of the interference $\imath(t)$.

The Signal-to-Interference at the beamformer output is therefore given by

$$\frac{S}{I} = \frac{p\mathbf{w}(\theta)^H \mathbf{a}(\theta)\mathbf{a}(\theta)^H \mathbf{w}(\theta)}{\mathbf{w}(\theta)^H \mathbf{R}_I \mathbf{w}(\theta)}. \tag{4.17}$$

Denoting the inner product between two vectors $\mathbf{u}$ and $\mathbf{v}$ as

$$(\mathbf{u}, \mathbf{v}) = \mathbf{u}^H \mathbf{R}_I \mathbf{v}, \tag{4.18}$$

we can rewrite (4.17) as

$$\frac{S}{I} = p\frac{\parallel (\mathbf{w}(\theta), \mathbf{R}_I{}^{-1}\mathbf{a}(\theta)) \parallel^2}{(\mathbf{w}(\theta), \mathbf{w}(\theta))}. \tag{4.19}$$

Now by the well-known Cauchy-Schwartz inequality this ratio is maximized for $\tilde{\mathbf{w}}(\theta)$ given by

$$\hat{\mathbf{w}}(\theta) = \beta \mathbf{R}_I{}^{-1}\mathbf{a}(\theta). \tag{4.20}$$

where β is some complex scalar.

To see the relation between (4.20) and (4.11), observe that if the signal and the interference are *uncorrelated* then the covariance matrix of $\mathbf{x}(t)$ is given by

$$\mathbf{R} = p\mathbf{a}(\theta)\mathbf{a}(\theta)^H + \mathbf{R}_I. \tag{4.21}$$

Hence, by the well-known matrix inversion lemma,

$$\mathbf{R}^{-1} = \mathbf{R}_I{}^{-1} - \mathbf{R}_I^{-1}\mathbf{a}(\theta)\left(\mathbf{a}(\theta)^H\mathbf{R}_I{}^{-1}\mathbf{a}(\theta) + \frac{1}{p}\right)^{-1}\mathbf{a}(\theta)^H\mathbf{R}_I{}^{-1}, \tag{4.22}$$

which implies, by multiplying (4.22) by $\mathbf{a}(\theta)$, that

$$\mathbf{R}^{-1}\mathbf{a}(\theta) = \alpha\mathbf{R}_I{}^{-1}\mathbf{a}(\theta), \tag{4.23}$$

where α is a complex scalar given by

$$\alpha = 1 - \left(\mathbf{a}(\theta)^H\mathbf{R}_I{}^{-1}\mathbf{a}(\theta) + \frac{1}{p}\right)^{-1}\mathbf{a}(\theta)^H\mathbf{R}_I{}^{-1}\mathbf{a}(\theta). \tag{4.24}$$

From (4.24) and (4.20) we get that

$$\hat{\mathbf{w}}(\theta) = \gamma\mathbf{R}^{-1}\mathbf{a}(\theta), \tag{4.25}$$

where γ is a scalar constant.

Hence, since $\hat{\mathbf{R}} \rightarrow \mathbf{R}$, the equivalence between the minimum variance weight vector and the maximal signal-to-interference weight vector, for the case of uncorrelated interferences, is established.

4.3 ADAPTED ANGULAR RESPONSE

An intersesting variant of the Minimum Variance solution have been proposed by Borgiotti and Kaplan (1979). Their starting point is the expression (4.11) for the optimal weight vector in the Minimum Variance solution. They are looking for a weight vector of the form

$$\mathbf{w}(\theta) = \mu \hat{\mathbf{R}}^{-1} \mathbf{a}(\theta). \tag{4.26}$$

However, unlike Capon, which determined μ so as to satisfy the constraint (4.7), they proposed to determine it so as to satisfy

$$\mathbf{w}(\theta)^H \mathbf{w}(\theta) = 1. \tag{4.27}$$

This modification leads to the desireable property that the contribution of the noise at the output of the linear combiner is, on the average, identical for every direction.

Combining (4.26) and (4.27), we get

$$\hat{\mu} = \frac{1}{\mathbf{a}(\theta)^H \hat{\mathbf{R}}^{-2} \mathbf{a}(\theta)}, \tag{4.28}$$

implying that the power output of the linear combiner, when steered towards direction θ is given by

$$A(\theta) = \frac{\mathbf{a}(\theta)^H \hat{\mathbf{R}}^{-1} \mathbf{a}(\theta)}{\mathbf{a}(\theta)^H \hat{\mathbf{R}}^{-2} \mathbf{a}(\theta)}. \tag{4.29}$$

Thus to solve our problem we have to first compute $A(\theta)$ for every θ, then determine the locations of the sources as the $\hat{q}$ peaks of $A(\theta)$ that pass a certain threshold, and finally steer beams to these directions to estimate the signal waveforms.

5. SUBOPTIMAL MODEL-BASED SOLUTIONS

Unlike the classical solutions presented in the previous section, in this section we present suboptimal solutions that exploit the structural and the statistical models of the signals and noise. Specifically, the solutions will heavily exploit the WN model. As for the signals model, though the derivation will be based on the SS model, the results are applicable also to the DS model. The only requirement is that the signals will not be coherent, i.e., that the covariance matrix of the signals will be *nonsingular*. Thus, the solutions we present are applicable to Non Coherent Signals and White Noise, which we refer to as NCSWN.

It should be pointed out that in the case of uniform linear array the requirement for nonsingularity of the covariance matrix can be overcomed by a preprocessing technique referred to as spatial smoothing, Evans *et. al.* (1982), Shan *et. al.* (1985).

Note that from (1.3) and the SS and WN models it follows that $\mathbf{x}(t)$ is a complex Gaussian vector with zero mean and covariance matrix given by

$$\mathbf{R} = \mathbf{A}(\Theta)\mathbf{C}\mathbf{A}(\Theta)^H + \sigma^2\mathbf{I}. \tag{5.1}$$

This matrix has interesting properties provided that the covariance matrix of the sources, $\mathbf{C}$, is *non singular*. Indeed, since $\mathbf{A}(\Theta)$ is full column-rank, the non-singularity of $\mathbf{C}$ implies that the $p \times p$ matrix $\mathbf{A}(\Theta)\mathbf{C}\mathbf{A}(\Theta)^H$ has rank q and hence that $p-q$ of its eigenvalues are zero. Consequently, denoting by $\lambda_1 \geq \ldots \geq \lambda_p$ and $\mathbf{v}_1, \ldots, \mathbf{v}_p$ the eigenvalues and the corresponding eigenvectors of $\mathbf{R}$, it follows from (5.1) that the $p-q$ smallest eigenvalues of $\mathbf{R}$ are all equal to σ^2, i.e.,

$$\lambda_{q+1} = \ldots = \lambda_p = \sigma^2, \tag{5.2}$$

and hence that the corresponding eigenvectors satisfy

$$\mathbf{R}\mathbf{v}_i = \sigma^2 \mathbf{v}_i \quad i = q+1, \ldots, p. \tag{5.3}$$

This implies, using (5.1), that

$$\mathbf{A}(\Theta)\mathbf{C}\mathbf{A}(\Theta)^H \mathbf{v}_i = \mathbf{0}. \quad i = q+1, \ldots, p, \tag{5.4}$$

and since $\mathbf{A}(\Theta)$ is full column rank and $\mathbf{C}$ is by assumption nonsingular, it then follows that

$$\mathbf{A}(\Theta)^H \mathbf{v}_i = \mathbf{0} \quad i = q+1, \ldots, p, \tag{5.5}$$

or alternatively,

$$\{\mathbf{a}(\theta_1), \ldots, \mathbf{a}(\theta_q)\} \perp \{\mathbf{v}_{q+1}, \ldots, \mathbf{v}_p\}. \tag{5.6}$$

Relations (5.2) and (5.6) are the key to the suboptimal solutions. Indeed, (5.2) implies that the number of sources can be determined from the multiplicity of the smallest eigenvalue, while (5.6) implies that the locations of the sources can be determined by searching over the array manifold for those steering vectors that are orthogonal to the eigensubspace of the smallest eigenvalue.

The problem is that in practice we do not have the eigenvalues and eigenvectors of the matrix $\mathbf{R}$ but only an estimate of them formed from the data. With probability one, the estimates do not obey, the relations (5.2) and (5.6). Instead, the eigenvalues corresponding to the smallest eigenvalue are "spread" around some value, and their corresponding eigenvectors are only "nearly orthogonal" to steering vectors of the sources.

As a result, to solve the detection and estimation problems we have to resort to more sophisticated approaches. To this end, it will be useful to have a compact representation of $\mathbf{R}$ in terms of its eigenstructure.

From (5.2) and the well known spectral representation theorem of matrix theory, we get

$$\mathbf{R} = \sum_{i=1}^{q} \lambda_i \mathbf{v}_i \mathbf{v}_i^H + \sum_{i=q+1}^{p} \sigma^2 \mathbf{v}_i \mathbf{v}_i^H . \tag{5.7}$$

Now since the eigenvectors form an orthonormal basis in $\mathbf{C}^P$, we have

$$\mathbf{I} = \sum_{i=1}^{q} \mathbf{v}_i \mathbf{v}_i^H + \sum_{i=q+1}^{p} \mathbf{v}_i \mathbf{v}_i^H , \tag{5.8}$$

which when substituted into (5.7) yields

$$\mathbf{R} = \sum_{i=1}^{q} (\lambda_i - \sigma^2) \mathbf{v}_i \mathbf{v}_i^H + \sigma^2 \mathbf{I}, \tag{5.9}$$

or in matrix notation,

$$\mathbf{R} = \mathbf{V}_S (\mathbf{\Lambda_S} - \sigma^2 \mathbf{I}) \mathbf{V}_S^H + \sigma^2 \mathbf{I}, \tag{5.10}$$

where

$$\mathbf{V}_S = [\mathbf{v}_1, \ldots, \mathbf{v}_q], \tag{5.11}$$

and

$$\mathbf{\Lambda_S} = \mathrm{diag}(\lambda_1, \ldots, \lambda_q). \tag{5.12}$$

5.1 ESTIMATOR OF THE EIGENSYSTEM

Having established the key role played by the eigensystem of $\mathbf{R}$, the first step is to estimate the eigensystem from the sampled data. To this end we derive the maximum likelihood estimator. The derivation follows Anderson (1963).

Relation (5.10) expresses $\mathbf{R}$ in terms of the eigensystem. The ML estimators of these parameters are the values that maximize the log-likelihood function given by

$$L(\mathbf{X}|\{\lambda_i\},\sigma^2,\{\mathbf{v}_i\}) = -M\log|\mathbf{R}| - M\mathrm{tr}[\mathbf{R}^{-1}\hat{\mathbf{R}}], \tag{5.13}$$

where $\hat{\mathbf{R}}$ is the sample-covariance matrix.

By applying the well-known matrix inversion lemma to (5.10) we get

$$\mathbf{R}^{-1} = \frac{1}{\sigma^2}\mathbf{I} - \mathbf{V}_S\mathbf{\Gamma_S}\mathbf{V}_S^H, \tag{5.14}$$

where

$$\mathbf{\Gamma_S} = \mathrm{diag}(\gamma_1,\ldots,\gamma_q) = \left[\left(\frac{1}{\sigma^2}\mathbf{\Lambda_S} - \sigma^2\mathbf{I}\right)^{-1} + \sigma^2\mathbf{I}\right]. \tag{5.15}$$

Also, by the well known spectral representation,

$$\hat{\mathbf{R}} = \mathbf{U}\mathbf{L}\mathbf{U}^H, \tag{5.16}$$

where

$$\mathbf{U} = [\mathbf{u}_1,\ldots,\mathbf{u}_p], \tag{5.17}$$

and

$$\mathbf{L} = \mathrm{diag}(l_1,\ldots,l_p). \tag{5.18}$$

Now, substituting (5.14) and (5.16) into (5.13), and using the fact that the determinant of a matrix is given by the product of its eigenvalues, we obtain

$$L(\mathbf{X}|\{\lambda_i\},\sigma^2,\{\mathbf{v}_i\}) = -M\sum_{i=1}^{q}\lambda_i - M(p-q)\log\sigma^2 - \frac{M}{\sigma^2}\mathrm{tr}[\hat{\mathbf{R}}] + M\mathrm{tr}[\mathbf{\Gamma_S}\mathbf{Q}^H\mathbf{L}\mathbf{Q}], \tag{5.19}$$

where $\mathbf{Q}$ is the $p \times q$ matrix

$$\mathbf{Q} = \mathbf{U}^H \mathbf{V}_S. \tag{5.20}$$

Note that in order to maximize this expression with respect to $\{\mathbf{v}_i\}$, it suffices to maximize with respect to all matrices $\mathbf{Q}$ such that $\mathbf{Q}^H\mathbf{Q} = \mathbf{I}$. To this end, note that since $\gamma_1 > \cdots > \gamma_q$ and $l_1 > \cdots l_p$, it can be readily verified that

$$\text{tr}[\mathbf{\Gamma_S}\mathbf{Q}^H\mathbf{L}\mathbf{Q}] \le \sum_{i=1}^{q} \gamma_i l_i, \tag{5.21}$$

with equality if and only if $\mathbf{Q}$ is given by

$$\hat{\mathbf{Q}} = \begin{pmatrix} \pm 1 & & & & 0 \\ & \cdot & & & \\ & & \cdot & & \\ & & & \cdot & \\ 0 & & & & \pm 1 \\ - & - & - & - & - \\ 0 & & & & 0 \end{pmatrix}. \tag{5.22}$$

Thus, the maximum of (5.19) will be achieved for $\mathbf{Q}$ given by (5.22). The ML estimate of the eigenvector matrix $\mathbf{V}_S$ is therefore given by

$$\hat{\mathbf{V}}_S = \mathbf{U}\hat{\mathbf{Q}}, \tag{5.23}$$

which implies that, up to a sign, the ML estimates of $\{\mathbf{v}_i\}$ are given by the corresponding eigenvectors of the sample-covariance matrix,

$$\hat{\mathbf{v}}_i = \mathbf{u}_i \qquad i = 1, \ldots, q. \tag{5.24}$$

Substituting the maximizing value of $\hat{\mathbf{Q}}$, from (5.22), into (5.19) and using the well known results that the determinant and the trace of a matrix are given by the product and sum, respectively, of its eigenvalues, we get

$$\begin{aligned} L(\mathbf{X}|\{\lambda_i\}, \sigma^2) &= -M\sum_{i=1}^{q} \log \lambda_i - M(p-q)\log \sigma^2 - \frac{M}{\sigma^2}\sum_{i=1}^{p} l_i + M\sum_{i=1}^{q}\left(\frac{l_i}{\sigma^2} - \frac{l_i}{\lambda_i}\right) \\ &= -M\sum_{i=1}^{q} \log \lambda_i - M(p-q)\log \sigma^2 - M\frac{1}{\sigma^2}\sum_{i=q+1}^{p} l_i - M\sum_{i=1}^{q}\frac{l_i}{\lambda_i} \end{aligned} \tag{5.25}$$

Now, by straightforward differentiation with respect to σ^2 we get that the ML estimator of this parameter is given by

$$\hat{\sigma}^2 = \frac{1}{p-q} \sum_{i=q+1}^{p} l_i. \tag{5.26}$$

This estimator is intuitively very pleasing. As one would expect from (5.2), the estimator of σ^2 is given by the average of the $p-q$ smallest eigenvalues of sample-covariance matrix.

Similarly, straightforward differentiation with respect to $\{\lambda_i\}$ yields

$$\hat{\lambda}_i = l_i \qquad i = 1, \ldots, q. \tag{5.27}$$

That is, the estimators of the large eigenvalues and their associated eigenvectors are given by their corresponding values in the sample-covariance matrix.

It should be pointed that since the eigenvectors corresponding to the smallest eigenvalue do not appear in the likelihood function, it follows that the only requirement on their ML estimators is that they span the orthogonal complement of the subspace spanned by the eigenvectors corresponding to the large eigenvalues. Therefore, up to a unitary transformation,

$$\hat{\mathbf{v}}_i = \mathbf{u}_i \qquad i = q+1, \ldots, p. \tag{5.28}$$

5.2 DETECTION OF THE NUMBER OF SOURCES

With the estimates of the eigenvalues at hand, we next address the problem of detecting the number of sources from the "multiplicity" of the smallest eigenvalue. The solution follows Wax and Kailath (1985) and is based on information theoretic criteria for model selection.

Let k denote the hypothesized number of signals, $k \in \{0, 1, \ldots, p-1\}$. With this notation, our model for the covariance matrix, from (5.10), can be rewritten as

$$\mathbf{R} = \sum_{i=1}^{k} (\lambda_i - \sigma^2)\mathbf{v}_i\mathbf{v}_i^H + \sigma^2\mathbf{I}. \tag{5.29}$$

Every k represents a different multiplicity and hence a different model. The problem is to determine the value of k that best fits the sampled data. We shall solve the problem by the MDL criterion (2.4).

Let $\phi^{(k)}$ denote the parameter vector of the k-th model,

$$\phi^{(k)} = (\lambda_1, \ldots, \lambda_k, \sigma^2, \mathbf{v}_1^T, \ldots, \mathbf{v}_k^T)^T. \tag{5.30}$$

To compute the first term in (2.4) we substitute the ML estimators of these parameters into the log-likelihood function (5.25), and add $M\sum_{i=1}^{p} \log l_i$, which is independent of k. This yields

$$L(\mathbf{X}|\hat{\phi}^{(k)}) = +M \sum_{i=k+1}^{p} \log l_i - M(p-k)\log\Big(\frac{1}{p-k}\sum_{i=k+1}^{p} l_i\Big), \tag{5.31}$$

or alternatively,

$$L(\mathbf{X}|\hat{\phi}^{(k)}) = -M(p-k)\log\left(\frac{\frac{1}{(p-k)}\sum_{i=k+1}^{p} l_i}{\prod_{i=k+1}^{p} l_i^{\frac{1}{(p-k)}}}\right). \tag{5.32}$$

We next compute the number of free parameters in $\phi^{(k)}$. Recall that a complex covariance matrix has real eigenvalues but complex eigenvectors. Thus the number of parameters in $\phi^{(k)}$ is $k+1+2pk$. However, since the eigenvectors are constrained to be mutually orthogonal and to have unit norm, their parameters are not independent. The mutual orthogonality constraints result in a reduction of $k(k-1)$ degrees of freedom, while the unit norm constraint results in a further reduction of k degrees of freedom. The number of degrees of freedom in $\phi^{(k)}$ is therefore $k+1+2pk-k-k(k-1) = k(2p-k+1)+1$.

Substituting (5.32) and the number of degrees-of-freedom into (2.4), the MDL criterion becomes

$$\hat{q} = \arg \min_{k \in \{0,\ldots,p-1\}} MDL_{NCSWN}(k), \tag{5.33}$$

where

$$MDL_{NCSWN}(k) = M(p-k)\log\left(\frac{\frac{1}{(p-k)}\sum_{i=k+1}^{p} l_i}{\left(\prod_{i=k+1}^{p} l_i\right)^{\frac{1}{(p-k)}}}\right) + \frac{1}{2}k(2p-k+1)\log M, \tag{5.34}$$

Analogously to the MDL criterion we can write down the AIC,

$$\hat{q} = \arg \min_{k \in \{0,\ldots,p-1\}} AIC_{NCSWN}(k), \tag{5.35}$$

where

$$AIC_{NCSWN}(k) = M(p-k)\log\left(\frac{\frac{1}{(p-k)}\sum_{i=k+1}^{p} l_i}{\left(\prod_{i=k+1}^{p} l_i\right)^{\frac{1}{(p-k)}}}\right) + k(2p-k+1). \tag{5.36}$$

5.3 ESTIMATION OF THE LOCATIONS

In the previous subsection we have seen how to estimate the number of sources from the eigenvalues of the sample-covariance matrix. Using this estimator, we now show how to estimate the locations of the sources from the eigenvectors of the sample-covariance matrix. We describe two different estimators. One is referred to as MUltiple SIgnal Classification (MUSIC) estimator and the other as Minimum Norm estimator.

5.3.1 The MUSIC Estimator

This estimator was proposed by Schmidt (1979) and Bienvenu and Kopp (1979, 1980).

As we have seen in (5.6), the steering vectors of the sources are orthogonal to the subspace spanned by the eigenvectors corresponding to the smallest eigenvalue of the covariance matrix. Thus if these eigenvectors were known, the locations of the sources could have been determined simply by searching over the array manifold for those steering vectors that are orthogonal to the subspace they span. However, in practice only an estimate of this subspace is available, given by the $\hat{q}$ eigenvectors corresponding to the smallest eigenvalues. Unfortunately, the problem is that this estimated subspace is not orthogonal, with probability one, to the steering vectors, thus rendering this simple approach unapplicable.

A simple solution to this problem is to introduce a "measure of orthogonality" and use it to select the $\hat{q}$ steering vectors which are "most nearly orthogonal" to the estimated subspace. A natural measure of orthogonality between a vector and a subspace is the squared cosine of the angle between this vector and the subspace. Since the eigenvectors $\mathbf{u}_{\hat{q}+1}, \ldots, \mathbf{u}_p$ are orthogonal, the squared cosine of the angle between the vector $\mathbf{a}(\theta)$ and the subspace span$\{\mathbf{u}_{\hat{q}+1}, \ldots, \mathbf{u}_p\}$, and hence the orthogonality measure, is given by

$$g(\theta) = \sum_{i=\hat{q}+1}^{p} \frac{\| \mathbf{a}(\theta)^H \mathbf{v}_i \|^2}{\mathbf{a}(\theta)^H \mathbf{a}(\theta)}. \tag{5.37}$$

The estimates of the locations of the sources are obtained by evaluating (5.37) for every θ and selecting the $\hat{q}$ values of θ that yield the lowest minima.

The asymptotic performance of this estimator, referred to as MUltiple Signal Classification (MUSIC), has been the topic of extensive investigation by Sharman *et al* (1984), Kaveh and Barbel (1986), Porat and Friedlander (1988) and Stoica and Nehorai (1989). The analysis shows that the estimator is consistent but that it is statistically efficient only in the special case that the covariance matrix of the sources $\mathbf{C}$ is diagonal. That is, when the sources are correlated, the MUSIC estimator does not achieve the Cramer-Rao lower bound.

5.3.2 The Minimum Norm Estimator

This estimator has been proposed by Kumaresan and Tufts (1984). In the special case of a uniform linear array this estimator coincides with the estimator proposed by Reddi (1979).

Instead of measuring the orthogonality between a vector and a subspace by the squared cosine of the angle between the vector and the subspace, as done in MUSIC, one can use the squared cosine of the angle between this vector and *some* vector $\mathbf{d}$ in this subspace. The only question is how to select $\mathbf{d}$.

To this end, observe that since $\mathbf{d}$ is in span$\{\mathbf{u}_{\hat{q}+1},\ldots,\mathbf{u}_p\}$, it is orthogonal to the complement subspace, that is

$$\mathbf{U}_{\hat{q}}^H \mathbf{d} = 0, \tag{5.38}$$

where $\mathbf{U}_S$ is the $p \times \hat{q}$ matrix of the signal subspace eigenvectors,

$$\mathbf{U}_S = [\mathbf{u}_1,\ldots,\mathbf{u}_{\hat{q}}]. \tag{5.39}$$

To determine $\mathbf{d}$ we can therefore fix the first component of $\mathbf{d}$ to be one and then solve the following set of linear equations

$$\tilde{\mathbf{V}}_S^H \tilde{\mathbf{d}} = \mathbf{g}^*, \tag{5.40}$$

where $*$ denotes the complex conjugate and $\mathbf{G}$, $\mathbf{g}$ and $\tilde{\mathbf{d}}$ are defined by

$$\mathbf{U}_S = \begin{bmatrix} \mathbf{g}^T \\ \mathbf{G} \end{bmatrix}, \tag{5.41}$$

and

$$\mathbf{d} = \begin{bmatrix} 1 \\ \tilde{\mathbf{d}} \end{bmatrix}. \tag{5.42}$$

Since this set of equations contains more unknown then equations, it does not have a unique solution. A question then arises as to which of all the possible solutions should be preferred. Motivated by the interpretation of the norm of $\tilde{\mathbf{d}}$ in the case of a uniform linear array, Kumaresan and Tufts proposed to select the minimum norm solution, given by

$$\tilde{\mathbf{d}} = (\mathbf{G}^H \mathbf{G})^{-1} \mathbf{G}^H \mathbf{g}^*. \tag{5.43}$$

A simpler form of the estimator can be obtained by exploiting the unitarity of the matrix $\mathbf{U}_S$. Indeed, we have

$$[\mathbf{g}^*, \mathbf{G}^H] \begin{bmatrix} \mathbf{g}^T \\ \mathbf{G} \end{bmatrix} = \mathbf{I}, \tag{5.44}$$

which implies that

$$\mathbf{g}^*\mathbf{g}^T + \mathbf{G}^H\mathbf{G} = \mathbf{I}, \tag{5.45}$$

and hence, by the matrix inversion lemma,

$$(\mathbf{G}^H\mathbf{G})^{-1} = (\mathbf{I} - \mathbf{g}^*\mathbf{g}^T)^{-1} = \mathbf{I} - \mathbf{g}^*(1 - \mathbf{g}^T\mathbf{g}^*)^{-1}\mathbf{g}^T. \tag{5.46}$$

Substituting this result into (5.45) and performing some algebraic manipulations, we get

$$\tilde{\mathbf{d}} = \frac{1}{1 - \mathbf{g}^H\mathbf{g}}\mathbf{G}\mathbf{g}^*. \tag{5.47}$$

We can express $\tilde{\mathbf{d}}$ also in terms of the noise subspace eigenvectors. To this end let $\mathbf{U}_N$ denote the matrix of the noise subspace eigenvectors,

$$\mathbf{U}_N = [\mathbf{u}_{\hat{q}+1}, \ldots, \mathbf{u}_p], \tag{5.48}$$

and let it be partioned as

$$\mathbf{U}_N = \begin{bmatrix} h^T \\ \mathbf{H} \end{bmatrix}. \tag{5.49}$$

Now since $[\mathbf{U}_S, \mathbf{U}_N]$ is an orthonormal basis in $\mathbf{C}^P$, it follows that

$$[\mathbf{U}_S, \mathbf{U}_N][\mathbf{U}_S, \mathbf{U}_N]^H = \mathbf{I}, \tag{5.50}$$

and hence

$$\begin{bmatrix} \mathbf{g}^T & \mathbf{h}^T \\ \mathbf{G} & \mathbf{H} \end{bmatrix} \begin{bmatrix} \mathbf{g}^* & \mathbf{G}^H \\ \mathbf{h}^* & \mathbf{H}^H \end{bmatrix} = \mathbf{I}. \tag{5.51}$$

Using this identity, the expression (5.47) becomes

$$\tilde{\mathbf{d}} = \frac{1}{\mathbf{h}^H\mathbf{h}}\mathbf{H}\mathbf{g}^*. \tag{5.52}$$

The vector $\mathbf{d}$, with $\tilde{\mathbf{d}}$ given by (5.47) or (5.52) "represents" the signal subspace. The squared cosine between $\mathbf{a}(\theta)$ and $\mathbf{d}$, which we use as a measure of orthogonality between $\mathbf{a}(\Theta)$ and the signal subspace, is given by

$$g(\theta) = \frac{\| \mathbf{a}(\theta)^H\mathbf{d} \|^2}{\mathbf{a}(\theta)^H\mathbf{a}(\theta)}. \tag{5.53}$$

The estimates of the locations of the sources are obtained by evaluating (5.53) for every θ and selecting the $\hat{q}$ values of θ that yield the lowest minima.

6. OPTIMAL SOLUTION FOR THE DSWN MODEL

In this section we present the optimal solution to the Deterministic Signals (DS) and White Noise (WN), which we refer to as the DSWN model.

Before we present this solution, we derive the maximum likelihood estimator which plays a key role in the solution.

6.1 THE MAXIMUM LIKELIHOOD ESTIMATOR

The derivation of Maximum Likelihood (ML) estimator follows Wax (1985).

From (1.3) and the DS and WN models, the joint density function of the sampled data is given by

$$f(\mathbf{X}) = \prod_{i=1}^{M} \frac{1}{\pi|\sigma^2\mathbf{I}|} \exp(-\frac{1}{\sigma^2} \| \mathbf{x}(t_i) - \mathbf{A}(\Theta)\mathbf{s}(t_i) \|^2). \tag{6.1}$$

In this statistical model the number of parameters grows linearly with the number of samples since $\{\mathbf{s}(t_i)\}$ represent $2qM$ free real parameters. This rather pathological behaviour of the DS model has serious implications on the performance of the ML estimator since, as is well known, the more unknowns there are the worse is the performance.

From (6.1), the log-likelihood of the sampled data, ignoring constant terms, is given by

$$L(\mathbf{X}|\Theta, \{\mathbf{s}(t_i)\}, \sigma^2) = -Mp\log\sigma^2 - \frac{1}{\sigma^2}\sum_{i=1}^{M} \| \mathbf{x}(t_i) - \mathbf{A}(\Theta)\mathbf{s}(t_i) \|^2 . \tag{6.2}$$

To compute the ML estimator we have to maximize the log-likelihood with respect to the unknown parameters. We carry out this maximization in three steps: first (i) maximize with respect to σ^2 with $\{\mathbf{s}(t_i)\}$ and Θ being fixed, then (ii) substitute the resulting estimate of σ^2 expressed as a function of $\{\mathbf{s}(t_i)\}$ and Θ, back into the log-likelihood function and maximize with respect to parameter $\{\mathbf{s}(t_i)\}$, with Θ fixed, and finally (iii) substitute the resulting estimate of $\{\mathbf{s}(t_i)\}$, expressed as a function of Θ, back into the log-likelihood function and obtain a function to be maximized over Θ only.

To carry out the first step, we fix Θ and $\{\mathbf{s}(t_i)\}$, and then maximize with respect to σ^2. This yields

$$\hat{\sigma}^2(\Theta, \{\mathbf{s}(t_i)\}) = \frac{1}{Mp} \sum_{i=1}^{M} \| \mathbf{x}(t_i) - \mathbf{A}(\Theta)\mathbf{s}(t_i) \|^2 . \tag{6.3}$$

Substituting this result into (6.2), the log-likelihood becomes

$$L(\mathbf{X}|\Theta, \{\mathbf{s}(t_i)\}) = -Mp \log\left(\sum_{i=1}^{M} \| \mathbf{x}(t_i) - \mathbf{A}(\Theta)\mathbf{s}(t_i) \|^2 \right). \tag{6.4}$$

Notice that since the logarithm is a monotonic function, the values of Θ and $\{\mathbf{s}(t_i)\}$ that maximize this expression are in fact the Least-Squares estimators and hence are meaningful even if the WN model is not valid.

To carry out the second step, we fix Θ and maximize this expression with respect to $\mathbf{s}(t_i)$. This yields

$$\hat{\mathbf{s}}(t_i) = \left[\mathbf{A}(\Theta)^H \mathbf{A}(\Theta)\right]^{-1} \mathbf{A}(\Theta)^H \mathbf{x}(t_i). \tag{6.5}$$

Substituting this expression into the log-likelihood function, ignoring constant terms, we get

$$L(\mathbf{X}|\Theta) = -Mp \log\left(\sum_{i=1}^{M} \| \mathbf{x}(t_i) - \mathbf{P}_{\mathbf{A}(\Theta)}\mathbf{x}(t_i) \|^2 \right), \tag{6.6}$$

where $\mathbf{P}_{\mathbf{A}(\Theta)}$ is the orthogonal projection onto the space spanned by the columns of the matrix $\mathbf{A}(\Theta)$

$$\mathbf{P}_{\mathbf{A}(\Theta)} = \mathbf{A}(\Theta)\left[\mathbf{A}(\Theta)^H \mathbf{A}(\Theta)\right]^{-1} \mathbf{A}(\Theta)^H. \tag{6.7}$$

Alternatively, we can rewrite this also as

$$L(\mathbf{X}|\Theta) = -Mp\log\left(\sum_{i=1}^{M} \| \mathbf{P}_{\mathbf{A}(\Theta)}^{\perp}\mathbf{x}(t_i) \|^2\right), \tag{6.8}$$

where $\mathbf{P}_{\mathbf{A}(\Theta)}^{\perp}$ is the orthogonal projection onto the orthogonal complement of the space spanned by the columns of the matrix $\mathbf{A}(\Theta)$,

$$\mathbf{P}_{\mathbf{A}(\Theta)}^{\perp} = \mathbf{I} - \mathbf{P}_{\mathbf{A}(\Theta)}. \tag{6.9}$$

Thus, the ML estimator $\hat{\Theta}$ is the solution of the following maximization problem:

$$\hat{\Theta} = \arg\max_{\Theta} \sum_{i=1}^{M} \| \mathbf{P}_{\mathbf{A}(\Theta)}\mathbf{x}(t_i) \|^2, \tag{6.10}$$

or, alternatively, the solution of the following minimization problem:

$$\hat{\Theta} = \arg\min_{\Theta} \sum_{i=1}^{M} \| \mathbf{P}_{\mathbf{A}(\Theta)}^{\perp}\mathbf{x}(t_i) \|^2 . \tag{6.11}$$

This estimator has an appealing geometric interpretation. Recall that in the the absence of noise the vector $\mathbf{x}(t)$ is confined to the q-dimensional space spanned by the columns of the matrix $\mathbf{A}(\Theta)$, referred to as the signal subspace. Expression (6.10) implies that the ML estimator selects the signal subspace which is "closest" to the sampled vectors $\{\mathbf{x}(t_i)\}$, where closeness is measured by the sum of squares of the projections of $\{\mathbf{x}(t_i)\}$ onto this subspace. Similarly, expression (6.11) implies that the ML estimator selects the noise subspace which is "mostly orthogonal" to the sampled vectors $\{\mathbf{x}(t_i)\}$, where the orthogonality is measured by the sum of squares of the projections of $\{\mathbf{x}(t_i)\}$ onto this subspace.

Since the projection operator is ideponent, i.e. $\mathbf{P}_{\mathbf{A}(\Theta)}\mathbf{P}_{\mathbf{A}(\Theta)} = \mathbf{P}_{\mathbf{A}(\Theta)}$, we have

$$\begin{aligned}\sum_{i=1}^{M} \| \mathbf{P}_{\mathbf{A}(\Theta)}\mathbf{x}(t_i) \|^2 &= \sum_{i=1}^{M} \mathbf{x}(t_i)^H \mathbf{P}_{\mathbf{A}(\Theta)}\mathbf{x}(t_i) \\ &= \sum_{i=1}^{M} \mathrm{tr}[\mathbf{P}_{\mathbf{A}(\Theta)}\mathbf{x}(t_i)\mathbf{x}(t_i)^H],\end{aligned} \tag{6.12}$$

where tr[ ] denotes the trace operator. Substituting this result into (6.10), we can recast the ML estimator as the solution to the following problems:

$$\max_{\Theta} \mathrm{tr}[\mathbf{P}_{\mathbf{A}(\Theta)}\hat{\mathbf{R}}], \tag{6.13}$$

or, alternatively,

$$\min_{\Theta} \mathrm{tr}[\mathbf{P}^{\perp}_{\mathbf{A}(\Theta)}\hat{\mathbf{R}}], \tag{6.14}$$

where $\hat{\mathbf{R}}$ is the sample covariance matrix

$$\hat{\mathbf{R}} = \frac{1}{M}\sum_{i=1}^{M} \mathbf{x}(t_i)\mathbf{x}(t_i)^H. \tag{6.15}$$

This expression shows the central role played by the sample-covariance matrix in this problem. Indeed, the sample-covariance is the ***sufficient statistic*** for the problem; all the relevant information in the data $\{\mathbf{x}(t_i)\}$ is captured by $\mathbf{R}$.

An interesting interpretation of the ML estimator is obtained by expressing (6.14) in terms of the eigenstructure of $\hat{\mathbf{R}}$. Let $l_1 \geq \ldots \geq l_p$ and $\mathbf{u}_1, \ldots, \mathbf{u}_p$ denote the eigenvalues and eigenvectors, respectively, of $\hat{\mathbf{R}}$. From the spectral representation theorem of matrix theory, we can express $\hat{\mathbf{R}}$ as

$$\hat{\mathbf{R}} = \sum_{i=1}^{p} l_i \mathbf{u}_i \mathbf{u}_i^H. \tag{6.16}$$

Using this representation and the properties of the projection and trace operators we can rewrite (6.13) as

$$\hat{\Theta} = \arg\max_{\Theta} \sum_{i=1}^{p} l_i \parallel \mathbf{P}_{\mathbf{A}(\Theta)} \mathbf{u}_i \parallel^2 . \tag{6.17}$$

This expression shows that the measure of "closeness" between the observed data, as summarized by the sufficient statistic $\hat{\mathbf{R}}$, and the potential signal subspace can be also expressed in terms of the projections of the eigenvectors of $\hat{\mathbf{R}}$ onto this subspace; the larger the eigenvalue the more important it is that the projection of the corresponding eigenvector onto the signal subspace be maximized.

It is of interest to examine the performance of the ML estimator as $M \to \infty$. Unfortunately, the classical results on the consistency and the statistical efficiency of the ML estimator cannot be used in this problem because the regularity conditions required for their applicability do not hold. This "pathological" behaviour stems from the linear growth of the number of unknown parameters with the number of samples we have pointed out above.

Following Stoica and Nehorai (1988), we first establish the consistency of the ML estimator. Assuming $\hat{\mathbf{R}} \to \mathbf{R}$ and $\frac{1}{M}\sum_{i=1}^{M} \mathbf{s}(t_i)\mathbf{s}(t_i)^H \to \mathbf{P}$ as $M \to \infty$, we get that as $M \to \infty$ the function to be minimized becomes

$$\mathrm{tr}[\mathbf{P}^{\perp}_{\mathbf{A}(\hat{\Theta})}\mathbf{R}] = \mathrm{tr}[\mathbf{P}^{\perp}_{\mathbf{A}(\hat{\Theta})}(\mathbf{A}(\Theta)\mathbf{P}\mathbf{A}(\Theta)] + \sigma^2\mathbf{I}) \geq (p-q)\sigma^2, \tag{6.18}$$

which is clearly minimized for $\hat{\Theta} = \Theta$. The consistency of $\hat{\Theta}$ does not imply, however, the consistency of the other parameters of the model. Indeed, as $M \to \infty$ we get

$$\hat{\sigma}^2 \to \frac{1}{p}\mathrm{tr}[\mathbf{P}^{\perp}_{\mathbf{A}(\Theta)}\mathbf{R}] = \frac{(p-q)}{p}\sigma^2, \tag{6.19}$$

and

$$\hat{\mathbf{s}}(t_i) \to [\mathbf{A}(\Theta)^H \mathbf{A}(\Theta)]^{-1} \mathbf{A}(\Theta)^H \mathbf{x}(t_i) = \mathbf{s}(t_i) + [\mathbf{A}(\Theta)^H \mathbf{A}(\Theta)]^{-1} \mathbf{A}(\Theta)^H \mathbf{n}(t_i). \tag{6.20}$$

The "pathological" behaviour of the ML estimator in the DS model is not reflected only in the inconsistency of the above estimators. As shown by Stoica and Nehorai (1988), the ML estimator of Θ is statistically inefficient, namely, it does not achieve the Cramer-Rao lower bound as $M \to \infty$.

6.2 SIMULTANEOUS DETECTION AND LOCALIZATION

We shall solve the detection and localization problems simultaneously via the MDL principle for model selection. The derivation follows Wax and Ziskind (1989b).

Denoting by k the unknown number of sources, our model (1.3) becomes

$$\mathbf{x}(t) = \mathbf{A}(\Theta^{(k)})\mathbf{s}(t) + \mathbf{n}(t), \tag{6.21}$$

where $\mathbf{A}(\Theta^{(k)})$ is a $p \times k$ matrix of the steering vectors of the sources,

$$\mathbf{A}(\Theta^{(k)}) = \left[\mathbf{a}(\theta_1), \ldots, \mathbf{a}(\theta_k)\right]. \tag{6.22}$$

Note that every different k represents a potential and competing model. Our problem is to choose the model in this family that best fits the sampled data.

Unfortunately, this family of models does not allow a straightforward application of the MDL principle for model selection since the number of parameters in this family grows with the number of samples. Indeed, as pointed in the derivation of the maximum likelihood estimator, the number of parameters in this family is $1 + q + 2kM$.

To overcome this problem, we shall recast the problem so as to eliminate the $2kM$ parameters of the signals $\{s(t_i)\}$, which are nuisance parameters in the detection and localization problems.

To this end, consider the partition of the vector $\mathbf{x}(t)$ into its components in the signal subspace, span$\{\mathbf{A}(\Theta^{(k)})\}$, and in its complement, the noise subspace. We have

$$\mathbf{x}(t) = \mathbf{G}(\Theta^{(k)}) \begin{bmatrix} \mathbf{x}_S(t) \\ \mathbf{x}_N(t) \end{bmatrix}, \tag{6.23}$$

where $\mathbf{x}_S(t)$ denotes the $k \times 1$ component in the signal subspace, $\mathbf{x}_N(t)$ denotes the $(p-k) \times 1$ component in the noise subspace, and $\mathbf{G}(\Theta^{(k)})$ denotes a unitary coordinate transformation matrix determined so as to allien the signal subspace with the first k coordinates, i.e.,

$$\mathbf{P}_{\mathbf{A}(\Theta^{(k)})}\mathbf{x}(t) = \mathbf{G}(\Theta^{(k)}) \begin{bmatrix} \mathbf{x}_S(t) \\ \mathbf{0} \end{bmatrix}. \tag{6.24}$$

Consequently, this transformation alliens the last $p-k$ coordinates in the noise subspace, i.e.,

$$\mathbf{P}^{\perp}_{\mathbf{A}(\Theta^{(k)})}\mathbf{x}(t) = \mathbf{G}(\Theta^{(k)}) \begin{bmatrix} \mathbf{0} \\ \mathbf{x}_N(t) \end{bmatrix}. \tag{6.25}$$

The representation (6.23) defines a family of competing models for the sampled vector $\mathbf{x}(t)$. To apply the MDL principle to this family we must compute the code length required to encode the data $\{\mathbf{x}(t_i)\}$ using each of the competing models. Now since $\mathbf{x}(t)$ is expresed in terms of its components in the signal and noise subspaces, the code length of $\{\mathbf{x}(t_i)\}$ should be expressed in these terms.

We start with the noise subspace components $\{\mathbf{x}_N(t_i)\}$. According to (2.4), the code length required to encode them depends on the model for their probability distribution. To this end, observe that from (6.21) and (6.25) we get

$$\mathbf{P}^{\perp}_{\mathbf{A}(\Theta^{(k)})}\mathbf{n}(t) = \mathbf{G}(\Theta^{(k)}) \begin{bmatrix} \mathbf{0} \\ \mathbf{x}_N(t) \end{bmatrix}. \tag{6.26}$$

Thus, since the noise $\mathbf{n}(t)$ is a complex Gaussian random vector with zero mean and covariance matrix $\sigma^2\mathbf{I}$, it follows that given $\Theta^{(k)}$, $\mathbf{x}_N(t)$ is a $(p-k)$-dimensional complex Gaussian random vector with zero mean and covariance matrix $\sigma^2\mathbf{I}$,

$$\mathbf{x}_N(t)|\Theta^{(k)} \sim N_{p-k}(\mathbf{0}, \sigma^2\mathbf{I}). \tag{6.27}$$

Hence the probabilistic model for the noise subspace component is given by

$$f(\{\mathbf{x}_N(t_i)\}|\Theta^{(k)}) = \prod_{i=1}^{M} \frac{1}{|\pi\sigma^2\mathbf{I}|}\exp\{-\mathbf{x}_N(t_i)^H\sigma^{-2}\mathbf{x}_N(t_i)\}, \tag{6.28}$$

which can be rewritten as

$$f(\{\mathbf{x}_N(t_i)\}|\Theta^{(k)}) = |\pi\sigma^2\mathbf{I}|^{-M}\exp\{-\sigma^{-2}M\mathrm{tr}[\hat{\mathbf{R}}_{NN}(\Theta^{(k)})]\}, \tag{6.29}$$

where $\hat{\mathbf{R}}_{NN}(\Theta^{(k)})$ denotes the $(p-k)\times(p-k)$ sample-covariance matrix of $\mathbf{x}_N(t)$,

$$\hat{\mathbf{R}}_{NN}(\Theta^{(k)}) = \frac{1}{M}\sum_{i=1}^{M}\mathbf{x}_N(t_i)\mathbf{x}_N(t_i)^H. \tag{6.30}$$

The probabilistic model (6.29) has only a single parameter, σ^2, whose ML estimator is given by

$$\hat{\sigma}^2(\Theta^{(k)}) = \frac{1}{(p-k)}\mathrm{tr}[\hat{\mathbf{R}}_{NN}(\Theta^{(k)})]. \tag{6.31}$$

Thus, it follows from (2.4) that the code length required to encode the noise subspace components, ignoring constant terms which are independent of k, is given by

$$L\{\mathbf{x}_N(t_i)|\Theta^{(k)}\} = M\log|\hat{\sigma}^2(\Theta^{(k)})\mathbf{I}| + M(p-k) + \frac{1}{2}\log M. \tag{6.32}$$

Next we evaluate the code length required to encode the signal subspace components $\{\mathbf{x}_S(t_i)\}$. To this end we have to select a probabilistic model for the conditional distribution of $\mathbf{x}_S(t)$ given $\mathbf{x}_N(t)$ and $\Theta^{(k)}$. A mathematically convenient choice is

$$\mathbf{x}_S(t)|\mathbf{x}_N(t), \Theta^{(k)} \sim N_k(\mathbf{B}\mathbf{x}_N(t), \mathbf{V}), \tag{6.33}$$

where $\mathbf{B}$ and $\mathbf{V}$ are unknown complex matrices of dimensions $k \times (p-k)$ and $k \times k$, respectively. The probabilistic model for the signal subspace components is therefore given by

$$f(\{\mathbf{x}_S(t_i)\}|\Theta^{(k)}, \{\mathbf{x}_N(t_i)\}) = \prod_{i=1}^{M} \frac{1}{|\pi\mathbf{V}|} \exp\{-(\mathbf{x}_S(t_i) - \mathbf{B}\mathbf{x}_N(t_i))^H \mathbf{V}^{-1}(\mathbf{x}_S(t_i) - \mathbf{B}\mathbf{x}_N(t_i))\}. \tag{6.34}$$

It can be easily verified that the number of free parameters in the matrices $\mathbf{B}$ and $\mathbf{V}$ is given by $2(p-k)k$ and k^2, respectively, and that their maximum likelihood estimators are given by

$$\hat{\mathbf{B}} = \hat{\mathbf{R}}_{SN}\hat{\mathbf{R}}_{NN}^{-1}, \tag{6.35}$$

$$\hat{\mathbf{V}} = \hat{\mathbf{R}}_{SS} - \hat{\mathbf{R}}_{SN}\hat{\mathbf{R}}_{NN}^{-1}\hat{\mathbf{R}}_{NS}, \tag{6.36}$$

where

$$\hat{\mathbf{R}}_{SS} = \frac{1}{M}\sum_{i=1}^{M} \mathbf{x}_S(t_i)\mathbf{x}_S(t_i)^H, \tag{6.37}$$

and

$$\hat{\mathbf{R}}_{SN} = \frac{1}{M}\sum_{i=1}^{M} \mathbf{x}_S(t_i)\mathbf{x}_N(t_i)^H. \tag{6.38}$$

From (2.4), the code length required to encode the signal subspace components is given by

$$L(\{\mathbf{x}_S(t_i)\}|\Theta^{(k)}, \{\mathbf{x}_N(t_i)\}) = M\log|\hat{\mathbf{V}}| + Mk + \frac{1}{2}(k^2 + 2k(p-k))\log M, \tag{6.39}$$

Combining (6.32) and (6.39), the total code length required to encode the signal and noise subspace components is given by

$$L(\{\mathbf{x}_N(t_i)\},\{\mathbf{x}_S(t_i)\}|\Theta^{(k)}) = M\log(\frac{1}{(p-k)}\mathrm{tr}\ \hat{\mathbf{R}}_{NN})^{p-k}$$
$$+ M\log|\hat{\mathbf{R}}_{SS} - \hat{\mathbf{R}}_{SN}\hat{\mathbf{R}}_{NN}^{-1}\hat{\mathbf{R}}_{NS}| + \frac{1}{2}(k^2+2k(p-k)+1)\log M. \quad (6.40)$$

In order to simplify this expression, we rewrite the sample-covariance matrix $\hat{\mathbf{R}}$, from (6.23), as

$$\hat{\mathbf{R}} = \mathbf{G}(\Theta^{(k)})\begin{pmatrix}\hat{\mathbf{R}}_{SS} & \hat{\mathbf{R}}_{SN}\\ \hat{\mathbf{R}}_{NS} & \hat{\mathbf{R}}_{NN}\end{pmatrix}\mathbf{G}(\Theta^{(k)})^H. \quad (6.41)$$

Taking the determinant of both sides, recalling that $\mathbf{G}(\Theta^{(k)})$ is unitary, we get

$$|\hat{\mathbf{R}}| = \begin{vmatrix}\hat{\mathbf{R}}_{SS} & \hat{\mathbf{R}}_{SN}\\ \hat{\mathbf{R}}_{NS} & \hat{\mathbf{R}}_{NN}\end{vmatrix} = |\hat{\mathbf{R}}_{NN}||\hat{\mathbf{R}}_{SS} - \hat{\mathbf{R}}_{SN}\hat{\mathbf{R}}_{NN}^{-1}\hat{\mathbf{R}}_{NS}|. \quad (6.42)$$

When substituted into (6.40), ignoring terms which are independent of k, this yields

$$L\{\mathbf{x}_N(t_i),\mathbf{x}_S(t_i)|\Theta^{(k)}\} = M(p-k)\log(\frac{\frac{1}{(p-k)}\mathrm{tr}\ \hat{\mathbf{R}}_{NN}(\Theta^{(k)})}{|\hat{\mathbf{R}}_{NN}(\Theta^{(k)})|^{\frac{1}{p-k}}}) + \frac{1}{2}k(2p-k)\log M. \quad (6.43)$$

Here we have changed the notation from $\hat{\mathbf{R}}_{NN}$ to $\hat{\mathbf{R}}_{NN}(\Theta^{(k)})$ to emphasize that $\hat{\mathbf{R}}_{NN}$ depends on $\Theta^{(k)}$. To make the dependence on Θ more explicit, observe that from (6.25) we get

$$\mathbf{P}^{\perp}_{\mathbf{A}(\Theta^{(k)})}\hat{\mathbf{R}}\mathbf{P}^{\perp}_{\mathbf{A}(\Theta^{(k)})} = \mathbf{G}(\Theta^{(k)})\begin{pmatrix}0 & 0\\ 0 & \hat{\mathbf{R}}_{NN}(\Theta^{(k)})\end{pmatrix}\mathbf{G}(\Theta^{(k)})^H. \quad (6.44)$$

Hence, by the invariance of the trace and the determinant under the transformation $\mathbf{F} \to \mathbf{G}\mathbf{F}\mathbf{G}^H$ when $\mathbf{G}$ is unitary,

$$\mathrm{tr}\ \hat{\mathbf{R}}_{NN}(\Theta^{(k)}) = \sum_{i=1}^{p-k} l_i(\Theta^{(k)}), \quad (6.45)$$

and

$$\det \hat{\mathbf{R}}_{NN}(\Theta^{(k)}) = \prod_{i=1}^{p-k} l_i(\Theta^{(k)}). \tag{6.46}$$

where $l_1(\Theta^{(k)}) \geq \ldots \geq l_{(p-k)}(\Theta^{(k)})$ denote the nonzero eigenvalues of the rank-$(p-k)$ matrix $\mathbf{P}^{\perp}_{\mathbf{A}(\Theta^{(k)})}\hat{\mathbf{R}}\mathbf{P}^{\perp}_{\mathbf{A}(\Theta^{(k)})}$.

Substituting (6.45) and (6.46) into (6.43), we get

$$L\{\mathbf{x}_N(t_i), \mathbf{x}_S(t_i)|\Theta^{(k)}\} = M(p-k)\log\left(\frac{\frac{1}{(p-k)}\sum_{i=1}^{p-k} l_i(\hat{\Theta}^{(k)})}{\left(\prod_{i=1}^{p-k} l_i(\hat{\Theta}^{(k)})\right)^{\frac{1}{(p-k)}}}\right) + \frac{1}{2}k(2p-k)\log M. \tag{6.47}$$

The code length (6.47) is conditioned on knowing the $k \times 1$ location vector $\Theta^{(k)}$. However, this value is unknown and hence it must be estimated from the data and encoded as a preamble to the code string. A natural choice for the estimator is the ML estimator derived above. Thus, substituting for Θ the ML estimator and increasing by k the number of free parameters in (6.43), the code length becomes

$$L(\mathbf{X}|k) = M(p-k)\log\left(\frac{\frac{1}{(p-k)}\sum_{i=1}^{p-k} l_i(\hat{\Theta}^{(k)})}{\prod_{i=1}^{p-k} l_i(\hat{\Theta}^{(k)})^{\frac{1}{(p-k)}}}\right) + \frac{1}{2}k(2p-k+1)\log M, \tag{6.48}$$

with $\hat{\Theta}^{(k)}$ being the ML estimator.

The MDL estimator of the number of sources is given by the value of k that minimizes the code length. Hence,

$$\hat{k} = \arg \min_{k \in \{0,\ldots,p-1\}} MDL_{DSWN}(k), \tag{6.49}$$

where

$$MDL_{DSWN}(k) = M(p-k)\log\left(\frac{\frac{1}{(p-k)}\sum_{i=1}^{p-k} l_i(\hat{\Theta}^{(k)})}{\prod_{i=1}^{p-k} l_i(\hat{\Theta}^{(k)})^{\frac{1}{(p-k)}}}\right) + \frac{1}{2}k(2p-k+1)\log M, \tag{6.50}$$

Analogously, the AIC criterion is given by

$$\hat{k} = \arg \min_{k \in \{0,\ldots,p-1\}} AIC_{DSWN}(k), \tag{6.51}$$

where

$$AIC_{DSWN}(k) = M(p-k)\log\Big(\frac{\frac{1}{(p-k)}\sum_{i=1}^{p-k} l_i(\hat{\Theta}^{(k)})}{\prod_{i=1}^{p-k} l_i(\hat{\Theta}^{(k)})^{\frac{1}{(p-k)}}}\Big) + k(2p-k+1). \tag{6.52}$$

6.3 BEAMFORMING

Having solved the detection and localization problem, the solution to the beamforming problem is striaghtforward. Indeed, from (6.5) we get

$$\hat{\mathbf{s}}(t_i) = \left[\mathbf{A}(\hat{\Theta}^{(\hat{q})})^H \mathbf{A}(\hat{\Theta}^{(\hat{q})})\right]^{-1} \mathbf{A}(\Theta^{(\hat{q})})^H \mathbf{x}(t_i). \tag{6.53}$$

7. OPTIMAL SOLUTION FOR THE SSWN MODEL

In this section we present the optimal solution for Stochastic Signals (SS) and White Noise (WN), which we refer to as the SSWN model.

Analogously to the DSWN model, the optimal solution involves a *simultaneous* solution to the detection and estimation problems and is based on the maximum likelihood estimator.

7.1 THE MAXIMUM LIKELIHOOD ESTIMATOR

The derivation of the ML estimator follows Bohme (1986) (see also Jaffer (1988)).

Note that from (1.3) and the SS and WN models it follows that $\mathbf{x}(t)$ is a complex Gaussian vector with zero mean and covariance matrix

$$\mathbf{R} = \mathbf{A}(\Theta)\mathbf{C}\mathbf{A}(\Theta)^H + \sigma^2\mathbf{I}. \tag{7.1}$$

Hence the joint density function of the sampled data is given by

$$f(\mathbf{X}) = \prod_{i=1}^{M} \frac{1}{\pi \det[\mathbf{R}]} \exp\{(-\mathbf{x}(t_i)^H \mathbf{R}^{-1}\mathbf{x}(t_i)\}, \tag{7.2}$$

Observe that unlike in the DSWN model, here the number of parameters does not grow with the number of samples. Indeed, the number of free parameters in σ^2, C and Θ is given by 1, q^2 and q, repectively.

The log-likelihood of the sampled data, ignoring constant terms, is given by

$$L(\mathbf{X}|\Theta, \mathbf{C}, \sigma^2) = -M \log|\mathbf{R}| - \sum_{i=1}^{M} \mathbf{x}(t_i)^H \mathbf{R}^{-1}\mathbf{x}(t_i), \tag{7.3}$$

or more compactly

$$L(\mathbf{X}|\Theta, \mathbf{C}, \sigma^2) = -M \log|\mathbf{R}| - M\mathrm{tr}[\mathbf{R}^{-1}\hat{\mathbf{R}}], \tag{7.4}$$

where $\hat{\mathbf{R}}$ is the sample-covariance matrix.

To compute the ML estimator we have to maximize the log-likelihood with respect to the unknown parameters. We shall carry out this maximization in three steps: first (i) maximize with respect to C with σ^2 and Θ being fixed, then (ii) substitute the resulting estimate of C, expressed as a function of σ^2 and Θ, back into the log-likelihood function and maximize with respect to parameter σ^2, with Θ fixed, and finally (iii) substitute the resulting estimate of σ^2, expressed as a function of Θ, back into the log-likelihood function and obtain a function to be maximized over Θ only.

To carry out the first step we have to differentiate (7.3) with respect to the matrix $\mathbf{C}$. Using the well known rules of matrix calculus, Graham (1983), and some rather lengthy algebraic manipulations, we get that the maximizing value of $\mathbf{C}$ expressed in terms of Θ and σ^2 is given by

$$\hat{\mathbf{C}}(\Theta,\sigma^2) = \left[\mathbf{A}(\Theta)^H\mathbf{A}(\Theta)\right]^{-1}\mathbf{A}(\Theta)^H(\hat{\mathbf{R}} - \sigma^2\mathbf{I})\mathbf{A}(\Theta)\left[\mathbf{A}(\Theta)^H\mathbf{A}(\Theta)\right]^{-1}. \tag{7.5}$$

Substituting this expression into (7.1), we get

$$\hat{\mathbf{R}}(\Theta,\sigma^2) = \left[\mathbf{A}(\Theta)^H\mathbf{A}(\Theta)\right]^{-1}\mathbf{A}(\Theta)^H\hat{\mathbf{R}}\mathbf{A}(\Theta)\left[\mathbf{A}(\Theta)^H\mathbf{A}(\Theta)\right]^{-1}\mathbf{A}(\Theta) + \sigma^2\mathbf{I}, \tag{7.6}$$

or more compactly, using the projection operator notation (6.7),

$$\hat{\mathbf{R}}(\Theta,\sigma^2) = \mathbf{P}_{\mathbf{A}(\Theta)}\hat{\mathbf{R}}\mathbf{P}_{\mathbf{A}(\Theta)} + \sigma^2\mathbf{P}^{\perp}_{\mathbf{A}(\Theta)}. \tag{7.7}$$

To carry out the second step in the computation of the ML estimator we substitute (7.7) into (7.4) and maximize it with respect to σ^2. After some straightforward algebraic manipulations, we get

$$\hat{\sigma}^2(\Theta) = \frac{1}{p-q}\mathrm{tr}\left[\mathbf{P}^{\perp}_{\mathbf{A}(\Theta)}\hat{\mathbf{R}}\right]. \tag{7.8}$$

Substituting this expression into (7.4), the log-likelikelihood becomes

$$L(\mathbf{X}|\Theta) = -M\log|\hat{\mathbf{R}}(\Theta)| - Mp, \tag{7.9}$$

where

$$\hat{\mathbf{R}}(\Theta) = \mathbf{P}_{\mathbf{A}(\Theta)}\hat{\mathbf{R}}\mathbf{P}_{\mathbf{A}(\Theta)} + \frac{1}{p-q}\mathrm{tr}\left[\mathbf{P}^{\perp}_{\mathbf{A}(\Theta)}\hat{\mathbf{R}}\right]\mathbf{P}^{\perp}_{\mathbf{A}(\Theta)}. \tag{7.10}$$

Hence, since the logarithm is a monotonic function, the ML estimator of Θ is given by the solution of the following minimization problem

$$\hat{\Theta} = \arg\min_{\Theta} |\hat{\mathbf{R}}(\Theta)|. \tag{7.11}$$

Though the intuitive interpretation of this estimator, in its present form, is rather obscure, we shall, in the sequel, present an intuitively pleasing interpretation of this estimator and a simple proof of its consistency.

Using the ML estimate of Θ, the ML estimates of σ^2 and $\mathbf{C}$ become

$$\hat{\sigma}^2 = \frac{1}{p-q}\mathrm{tr}\big[\mathbf{P}^{\perp}_{\mathbf{A}(\hat{\Theta})}\hat{\mathbf{R}}\big]. \tag{7.12}$$

and

$$\hat{\mathbf{C}} = \big[\mathbf{A}(\hat{\Theta})^H\mathbf{A}(\hat{\Theta})\big]^{-1}\mathbf{A}(\hat{\Theta})^H\big[\hat{\mathbf{R}} - \hat{\sigma}^2\mathbf{I}\big]\mathbf{A}(\hat{\Theta})\big[\mathbf{A}(\hat{\Theta})^H\mathbf{A}(\hat{\Theta})\big]^{-1}. \tag{7.13}$$

It can be readily verified that unlike in the DSWN model, the ML estimators of σ^2 and $\mathbf{C}$ are consistent. Indeed, assuming that $\hat{\Theta}$ is consistent, as $M \to \infty$ we get

$$\hat{\sigma}^2 \to \frac{1}{p-q}\mathrm{tr}\big[\mathbf{P}^{\perp}_{\mathbf{A}(\Theta)}\mathbf{R}\big] = \frac{\sigma^2}{p-q}\mathrm{tr}\big[\mathbf{P}^{\perp}_{\mathbf{A}}\big] = \sigma^2, \tag{7.14}$$

and

$$\hat{\mathbf{C}} \to \big[\mathbf{A}(\Theta)^H\mathbf{A}(\Theta)\big]^{-1}\mathbf{A}(\Theta)^H\big[\mathbf{R} - \sigma^2\mathbf{I}\big]\mathbf{A}(\Theta)\big[\mathbf{A}(\Theta)^H\mathbf{A}(\Theta)\big]^{-1} = \mathbf{C}. \tag{7.15}$$

Observe that $\hat{\mathbf{C}}$ is a "sensible" estimate of $\mathbf{C}$. Indeed, $\hat{\mathbf{C}}$ is obtained by simply solving (7.1) for $\mathbf{C}$, with the unknown quantities replaced by their estimates from the data. In fact, Schmidt (1979) proposed this estimator on intuitive grounds, only in his approach $\hat{\Theta}$ and $\hat{\sigma}^2$ were not the ML estimators. Interestingly, note that $\hat{\mathbf{C}}$ is not guranteed to be non-negative definite for finite sample sizes.

7.2 SIMULTANEOUS DETECTION AND LOCALIZATION

Since the number of parameters in the SSWN model does not grow with the number of samples, the simultaneous solution to the detection and localization problems can be derived by straightforwardly applying the MDL criterion for model selection in conjunction with the ML estimator derived above. Nevertheless, we present a different approach, based also on the MDL principle, which exploits the structure of the signal and noise subspaces and yields an intuitively pleasing interpretation of the ML estimator. The derivation follows Wax (1989a).

Analogously to the DSWN model, we shall evaluate the description length in three steps: (i) compute the description length of the noise subspace components $\{\mathbf{x}_N(t_i)\}$ assuming that $\Theta^{(k)}$ is given, (ii) compute the description length of the signal subspace components $\{\mathbf{x}_S(t_i)\}$ assuming that the noise subspace components $\{\mathbf{x}_N(t_i)\}$ and $\Theta^{(k)}$ are given, and (iii) compute the description length of $\Theta^{(k)}$.

Observe that since the noise model is identical to that used in the DSWN model, the description length of the noise subspace components is identical to that given in (6.32). Nevertheless, we rederive this expression here for sake of completeness.

The WN model and (6.26) imply that $\mathbf{x}_N(t)$ is a $(p-k)\times 1$ complex Gaussian random vector with zero mean and covariance matrix $\sigma^2\mathbf{I}$, i.e.,

$$\mathbf{x}_N(t)|\Theta^{(k)} \sim N_{p-k}(\mathbf{0},\sigma^2\mathbf{I}). \tag{7.16}$$

Hence the probabilistic model for the noise subspace components is given by

$$f(\{\mathbf{x}_N(t_i)\}|\Theta^{(k)}) = \prod_{i=1}^{M} \frac{1}{|\pi\sigma^2\mathbf{I}|}\exp\{-\mathbf{x}_N^H(t_i)\sigma^{-2}\mathbf{x}_N(t_i)\}, \tag{7.17}$$

which can be rewritten as

$$f(\{\mathbf{x}_N(t_i)\}|\Theta^{(k)}) = |\pi\sigma^2\mathbf{I}|^{-M}\exp\{-\sigma^{-2}M\mathrm{tr}[\hat{\mathbf{R}}_{NN}(\Theta^{(k)})]\}, \tag{7.18}$$

where $\hat{\mathbf{R}}_{NN}(\Theta^{(k)})$ denotes the $(p-k)\times(p-k)$ sample-covariance matrix of $\mathbf{x}_N(t)$,

$$\hat{\mathbf{R}}_{NN}(\Theta^{(k)}) = \frac{1}{M}\sum_{i=1}^{M}\mathbf{x}_N(t_i)\mathbf{x}_N^H(t_i). \tag{7.19}$$

As can be readily verified, the probabilistic model (7.18) has only a single parameter, σ^2, and its ML estimator is given by

$$\hat{\sigma}^2(\Theta^{(k)}) = \frac{1}{(p-k)}\mathrm{tr}[\hat{\mathbf{R}}_{NN}(\Theta^{(k)})]. \tag{7.20}$$

It then follows from (2.16) that the code length required to encode the noise subspace components, ignoring constant terms which are independent of k, is given by

$$L\{\mathbf{x}_N(t_i)|\Theta^{(k)}\} = M\log|\hat{\sigma}^2(\Theta^{(k)})\mathbf{I}| + M(p-k) + \frac{1}{2}\log M. \tag{7.21}$$

Next, we evaluate the description length of the signal subspace components $\{\mathbf{x}_S(t_i)\}$. Note that the SS model and (6.24) imply that $\mathbf{x}_S(t)$ is a $k\times 1$ complex Gaussian random vector with zero mean. Denoting by $\mathbf{R}_{SS}(\Theta^{(k)})$ its covariance matrix, we have

$$\mathbf{x}_S(t)|\Theta^{(k)} \sim N_k(\mathbf{0},\mathbf{R}_{SS}(\Theta^{(k)})). \tag{7.22}$$

The probabilistic model for the signal subspace components is therefore given by

$$f(\{\mathbf{x}_S(t_i)\}|\Theta^{(k)}) = \prod_{i=1}^{M}\frac{1}{|\pi\mathbf{R}_{SS}(\Theta^{(k)})|}\exp\{-\mathbf{x}_S^H(t_i)\mathbf{R}_{SS}^{-1}(\Theta^{(k)})\mathbf{x}_S(t_i)\}, \tag{7.23}$$

which can be rewritten as

$$f(\{\mathbf{x}_S(t_i)\}|\Theta^{(k)}) = |\pi \mathbf{R}_{SS}(\Theta^{(k)})|^{-M} \exp\{-M \text{tr}[\mathbf{R}_{SS}^{-1}(\Theta^{(k)})\hat{\mathbf{R}}_{SS}(\Theta^{(k)})]\}, \tag{7.24}$$

where

$$\hat{\mathbf{R}}_{SS}(\Theta^{(k)}) = \frac{1}{M}\sum_{i=1}^{M} \mathbf{x}_S(t_i)\mathbf{x}_S^H(t_i). \tag{7.25}$$

As can be readily verified, the number of free real parameters in $\mathbf{R}_{SS}(\Theta^{(k)})$ is k^2 and its maximum likelihood estimator is given by the sample covariance matrix $\hat{\mathbf{R}}_{SS}(\Theta^{(k)})$. Consequntly, from (2.16), the description length of the signal subspace components, ignoring constant terms which are independent of k, is given by

$$L\{\mathbf{x}_S(t_i)|\Theta^{(k)}\} = M \log|\hat{\mathbf{R}}_{SS}(\Theta^{(k)})| + Mk + \frac{1}{2}k^2 \log M. \tag{7.26}$$

Thus, summing up (7.21) and (7.26), again ignoring constant terms which are independent of k, the total description length of the signal and noise subspace components is given by

$$L\{\mathbf{x}_N(t_i), \mathbf{x}_S(t_i)|\Theta^{(k)}\} = M \log(|\hat{\mathbf{R}}_{SS}(\Theta^{(k)})||\hat{\sigma}^2(\Theta^{(k)})\mathbf{I}|) + \frac{1}{2}(k^2+1)\log M. \tag{7.27}$$

The computed description length was conditioned on knowing the $k \times 1$ location vector $\Theta^{(k)}$. However, this value is unknown and hence it must be estimated from the data and encoded as a preamble to the code string. Now, since our goal is to obtain the shortest code length, which we claim assures the best detection performance, the optimal estimate is obtained by minimizing (7.27), i.e.,

$$\hat{\Theta}^{(k)} = \arg\min_{\Theta^{(k)}} \log(|\hat{\mathbf{R}}_{SS}(\Theta^{(k)})||\hat{\sigma}^2(\Theta^{(k)})\mathbf{I}|). \tag{7.28}$$

This estimator has an interesting geometric interpretation. Indeed, note that since $\mathbf{R}_{SS}(\Theta^{(k)}) = \frac{1}{M}\mathbf{X}_S\mathbf{X}_S^H$, where $\mathbf{X}_S = [\mathbf{x}_S(t_1), \ldots, \mathbf{x}_S(t_M)]$, it follows that $|\hat{\mathbf{R}}_{SS}(\Theta^{(k)})|$ is a Grammian and as such it represents the volume occupied by the signal subspace components. Note also that since $\sigma^2(\Theta^{(k)})$ represents the variance of each of the $(p-k)$ noise subspace components, it follows that $|\hat{\sigma}^2(\Theta^{(k)})\mathbf{I}|$ represents the volume of the noise subspace components under the spherical WN model. We can therefore interperate the estimator (7.28) as the value of $\Theta^{(k)}$ that minimizes the volume occupied by the data in *both* the signal and the noise subspaces. For comparison, the ML estimator of the deterministic signals model, (6.14), minimizes the value of $\hat{\sigma}^2 = \frac{1}{p-k}\text{tr}[\mathbf{P}^{\perp}_{\mathbf{A}(\hat{\Theta})}\hat{\mathbf{R}}]$ which is equivalent to minimizing the volume $|\hat{\sigma}^2(\Theta^{(k)})\mathbf{I}|$.

As we shall now show, the estimator (7.28) coincides with the ML estimator (7.11). To establish this, notice that from (6.24) and (6.25), suppressing the index k for notational compactness, we get

$$\mathbf{P}_{\mathbf{A}(\Theta)}\hat{\mathbf{R}}\mathbf{P}_{\mathbf{A}(\Theta)} = \mathbf{G}(\Theta)\begin{pmatrix} \hat{\mathbf{R}}_{SS}(\Theta) & \mathbf{0} \\ \mathbf{0} & \mathbf{0} \end{pmatrix}\mathbf{G}^H(\Theta), \tag{7.29}$$

and

$$\mathbf{P}^{\perp}_{\mathbf{A}(\Theta)}\hat{\mathbf{R}}\mathbf{P}^{\perp}_{\mathbf{A}(\Theta)} = \mathbf{G}(\Theta)\begin{pmatrix} \mathbf{0} & \mathbf{0} \\ \mathbf{0} & \hat{\mathbf{R}}_{NN}(\Theta) \end{pmatrix}\mathbf{G}^H(\Theta), \tag{7.30}$$

where $\hat{\mathbf{R}}$ denotes the sample-covariance matrix, and also

$$\mathbf{P}^{\perp}_{\mathbf{A}(\Theta)}\hat{\sigma}^2(\Theta)\mathbf{P}^{\perp}_{\mathbf{A}(\Theta)} = \mathbf{G}(\Theta)\begin{pmatrix} \mathbf{0} & \mathbf{0} \\ \mathbf{0} & \hat{\sigma}^2(\Theta)\mathbf{I} \end{pmatrix}\mathbf{G}^H(\Theta). \tag{7.31}$$

Now, taking the trace of both sides of (7.30), recalling that $\mathbf{G}(\Theta^{(k)})$ is unitary, we have

$$\text{tr}\,[\hat{\mathbf{R}}_{NN}(\Theta)] = \text{tr}\,[\mathbf{P}^{\perp}_{\mathbf{A}(\Theta)}\hat{\mathbf{R}}], \tag{7.32}$$

and hence, by (7.20),

$$\hat{\sigma}^2(\Theta^{(k)}) = \frac{1}{p-k}\text{tr}\,[\mathbf{P}^{\perp}_{\mathbf{A}(\Theta^{(k)})}\hat{\mathbf{R}}]. \tag{7.33}$$

Also, summing up (7.29) and (7.31) we get

$$\mathbf{P}_{\mathbf{A}(\Theta)}\hat{\mathbf{R}}\mathbf{P}_{\mathbf{A}(\Theta)} + \mathbf{P}^{\perp}_{\mathbf{A}(\Theta)}\hat{\sigma}^2(\Theta) = \mathbf{G}(\Theta)\begin{pmatrix}\hat{\mathbf{R}}_{SS}(\Theta) & 0 \\ 0 & \hat{\sigma}^2(\Theta)\mathbf{I}\end{pmatrix}\mathbf{G}^H(\Theta). \tag{7.34}$$

Taking the determinant of both sides, recalling that $\mathbf{G}(\Theta)$ is unitary, we get

$$|\mathbf{P}_{\mathbf{A}(\Theta)}\hat{\mathbf{R}}\mathbf{P}_{\mathbf{A}(\Theta)} + \mathbf{P}^{\perp}_{\mathbf{A}(\Theta)}\hat{\sigma}^2(\Theta)| = |\hat{\mathbf{R}}_{SS}(\Theta)||\hat{\sigma}^2(\Theta)\mathbf{I}|, \tag{7.35}$$

and therefore, by (7.28),

$$\hat{\Theta}^{(k)} = \arg\min_{\Theta^{(k)}} |\mathbf{P}_{\mathbf{A}(\Theta^{(k)})}\hat{\mathbf{R}}\mathbf{P}_{\mathbf{A}(\Theta^{(k)})} + \frac{1}{p-k}\text{tr}\,[\mathbf{P}^{\perp}_{\mathbf{A}(\Theta^{(k)})}\hat{\mathbf{R}}]\mathbf{P}^{\perp}_{\mathbf{A}(\Theta^{(k)})}|, \tag{7.36}$$

which coincides with the ML estimator (7.11).

Having shown that the estimator (7.28) coincides with the ML estimator, we shall now prove the consistency of the ML estimator using this more convenient form.

From (6.23), suppressing the index k, we get

$$\hat{\mathbf{R}} = \mathbf{G}(\Theta)\begin{pmatrix}\hat{\mathbf{R}}_{SS}(\Theta) & \hat{\mathbf{R}}_{SN}(\Theta) \\ \hat{\mathbf{R}}_{NS}(\Theta) & \hat{\mathbf{R}}_{NN}(\Theta)\end{pmatrix}\mathbf{G}^H(\Theta), \tag{7.37}$$

where $\hat{\mathbf{R}}_{SS}(\Theta)$ and $\hat{\mathbf{R}}_{NN}(\Theta)$ are given by (7.25) and (7.19), respectively, and

$$\hat{\mathbf{R}}_{SN}(\Theta) = \frac{1}{M}\sum_{i=1}^{M}\mathbf{x}_S(t_i)\mathbf{x}_N^H(t_i). \tag{7.38}$$

Taking the determinant of both sides, recalling that $\mathbf{G}(\Theta)$ is unitary, we get

$$|\hat{\mathbf{R}}| = \begin{vmatrix}\hat{\mathbf{R}}_{SS}(\Theta) & \hat{\mathbf{R}}_{SN}(\Theta) \\ \hat{\mathbf{R}}_{NS}(\Theta) & \hat{\mathbf{R}}_{NN}(\Theta)\end{vmatrix} = |\hat{\mathbf{R}}_{NN}(\Theta)||\hat{\mathbf{R}}_{SS}(\Theta) - \hat{\mathbf{R}}_{SN}(\Theta)\hat{\mathbf{R}}^{-1}_{NN}(\Theta)\hat{\mathbf{R}}_{NS}(\Theta)|, \tag{7.39}$$

which, since $\hat{\mathbf{R}}_{SN}(\Theta)\hat{\mathbf{R}}_{NN}^{-1}(\Theta)\hat{\mathbf{R}}_{NS}(\Theta)$ is non-negative definite, implies that

$$|\hat{\mathbf{R}}| \leq |\hat{\mathbf{R}}_{NN}(\Theta)||\hat{\mathbf{R}}_{SS}(\Theta)|. \tag{7.40}$$

Now

$$|\hat{\mathbf{R}}_{NN}(\Theta)| = \prod_{i=1}^{p-q} l_i^N(\Theta), \tag{7.41}$$

and by (7.33)

$$|\hat{\sigma}^2(\Theta)\mathbf{I}| = \left(\frac{1}{p-q}\sum_{i=1}^{p-q} l_i^N(\Theta)\right)^{p-q}, \tag{7.42}$$

where $l_1^N(\Theta) \geq \ldots \geq l_{p-q}^N(\Theta)$ denote the nonzero eigenvalues of the rank-$(p-q)$ matrix $\mathbf{P}_{\mathbf{A}(\Theta)}^{\perp}\hat{\mathbf{R}}\mathbf{P}_{\mathbf{A}(\Theta)}^{\perp}$.

By the inequality of the arithmetic and geometric means we have

$$\frac{1}{p-q}\sum_{i=1}^{p-q} l_i^N(\Theta) \geq \left(\prod_{i=1}^{p-q} l_i^N(\Theta)\right)^{\frac{1}{p-q}}, \tag{7.43}$$

with equality if and only if $l_1^N(\Theta) = \ldots = l_{p-q}^N(\Theta)$. Hence, using (7.41) and (7.42), we obtain

$$|\hat{\mathbf{R}}_{NN}(\Theta)| \leq |\hat{\sigma}^2(\Theta)\mathbf{I}|, \tag{7.44}$$

and therefore, by (7.40),

$$|\hat{\mathbf{R}}| \leq |\hat{\mathbf{R}}_{SS}(\Theta)||\hat{\sigma}^2(\Theta)\mathbf{I}|, \tag{7.45}$$

with equality if and only if $\hat{\mathbf{R}}_{NS}(\Theta) = \mathbf{0}$ and $l_1^N(\Theta) = \ldots = l_{p-q}^N(\Theta)$.

Since both conditions are satisfied for the true Θ as $M \to \infty$, it follows that the minimum of (7.28) is obtained at the true Θ. This establishes the consistency of the ML estimator $\hat{\Theta}$.

Yet another form of the ML estimator can be obtained by casting it in terms of the eigenvalues of the matrices involved. Indeed, from (7.29) and (7.30), by the well known properties of the trace and the determinant and their invariance under the transformation $\mathbf{F} \rightarrow \mathbf{G}\mathbf{F}\mathbf{G}^H$ when $\mathbf{G}$ is unitary, we get

$$|\hat{\mathbf{R}}_{SS}(\Theta^{(k)})| = \prod_{i=1}^{k} l_i^S(\Theta^{(k)}), \tag{7.46}$$

and

$$\text{tr }[\hat{\mathbf{R}}_{NN}(\Theta^{(k)})] = \sum_{i=1}^{p-k} l_i^N(\Theta^{(k)}), \tag{7.47}$$

where $l_1^S(\Theta^{(k)}) \geq \ldots \geq l_k^S(\Theta^{(k)})$ denote the nonzero eigenvalues of the rank-k matrix $\mathbf{P}_{\mathbf{A}(\Theta^{(k)})}\hat{\mathbf{R}}\mathbf{P}_{\mathbf{A}(\Theta^{(k)})}$ and $l_1^N(\Theta^{(k)}) \geq \ldots \geq l_{p-k}^N(\Theta^{(k)})$ denote the nonzero eigenvalues of the rank-$(p-k)$ matrix $\mathbf{P}^{\perp}_{\mathbf{A}(\Theta^{(k)})}\hat{\mathbf{R}}\mathbf{P}^{\perp}_{\mathbf{A}(\Theta^{(k)})}$.

From (7.20) we therefore obtain

$$\hat{\sigma}^2(\Theta) = \frac{1}{p-k}\sum_{i=1}^{p-k} l_i^N(\Theta^{(k)}), \tag{7.48}$$

and hence, substituting (7.48) and (7.46) into (7.28), we get

$$\hat{\Theta}^{(k)} = \arg\min_{\Theta^{(k)}}\Big(\prod_{i=1}^{k} l_i^S(\Theta^{(k)})\Big)\Big(\frac{1}{p-k}\sum_{i=1}^{p-k} l_i^N(\Theta^{(k)})\Big)^{p-k}. \tag{7.49}$$

Substituting the ML estimator $\hat{\Theta}^{(k)}$ into (7.28) and adding $\frac{1}{2}k \log M$ for the description length of the k real parameters, we get that the MDL estimator of the number of sources is given by the value of k which minimizes the following criterion:

$$\hat{k} = \arg\min_{k} MDL_{SSWN}(k), \tag{7.50}$$

where

$$MDL_{SSWN}(k) = M\log\Big(\prod_{i=1}^{k} l_i^S(\hat{\Theta}^{(k)})\Big(\frac{1}{p-k}\sum_{i=1}^{p-k} l_i^N(\hat{\Theta}^{(k)})\Big)^{p-k}\Big) + \frac{1}{2}k(k+1)\log M, \tag{7.51}$$

with $\hat{\Theta}^{(k)}$ denoting the ML estimator (7.28).

Similarly, the AIC estimator is given by

$$\hat{k} = \arg\min_{k} AIC_{SSWN}(k), \tag{7.52}$$

where

$$AIC_{SSWN}(k) = M \log\Big(\prod_{i=1}^{k} l_i^S(\hat{\Theta}^{(k)})\big(\frac{1}{p-k}\sum_{i=1}^{p-k} l_i^N(\hat{\Theta}^{(k)})\big)^{p-k}\Big) + k(k+1). \tag{7.53}$$

7.3 BEAMFORMING

Unlike in the DSWN model, in the SSWN model the signals $\{s(t_i)\}$ are not parameters of the model and hence the solution of the beamforming problem is less striaghtforward. The presentation follows Wax (1985).

Note that from (1.3) and the SS model it follows that if Θ is given then $s(t_i)$ is independent of $x(t_j)$ for $j \neq i$. Consequently, having an estimate of Θ at hand, it is natural to base the estimate of $s(t_i)$ solely on $x(t_i)$.

To derive the estimate, observe that the SSWN model implies that $s(t_i)$ and $x(t_i)$ are jointly Gaussian. This, in turn, implies that the best estimator of $s(t_i)$ given $x(t_i)$ is the conditional mean $E[s(t_i)|x(t_i)]$, where E denotes the expectation operator, given by the well known expression

$$\hat{s}(t_i) = E[s(t_i)x(t_i)^H]E[x(t_i)x(t_i)^H]^{-1}x(t_i). \tag{7.54}$$

Now, from (1.3), we have

$$E[s(t_i)x(t_i)^H] = E[s(t_i)s(t_i)^H]\mathbf{A}(\Theta)^H = \mathbf{CA}(\Theta)^H, \tag{7.55}$$

so that we can rewrite (7.54) as

$$\hat{s}(t_i) = \mathbf{CA}(\Theta)^H\mathbf{R}^{-1}x(t_i). \tag{7.56}$$

Replacing the above quantities by their estimates from the data, we get

$$\hat{\mathbf{s}}(t_i) = \hat{\mathbf{C}}^{(\hat{q})}\mathbf{A}(\hat{\Theta}^{(\hat{q})})^H\hat{\mathbf{R}}^{-1}\mathbf{x}(t_i). \tag{7.57}$$

Since $\mathbf{R}$ is given by (7.1), and since the following easily verified matrix identity holds,

$$\mathbf{AC}(\mathbf{ACA}^H + \sigma^2\mathbf{I})^{-1} = (\mathbf{A}^H\mathbf{A} + \sigma^2\mathbf{C}^{-1})^{-1}\mathbf{A}^H, \tag{7.58}$$

we get

$$\hat{\mathbf{s}}(t_i) = \left(\mathbf{A}(\hat{\Theta}^{(\hat{q})})^H\mathbf{A}(\hat{\Theta}^{(\hat{q})}) + \hat{\sigma}^2(\hat{\mathbf{C}}^{(\hat{q})})^{-1}\right)^{-1}\mathbf{A}(\hat{\Theta}^{(\hat{q})})^H\mathbf{x}(t_i). \tag{7.59}$$

The asymptotic behaviour of (7.59) can be easily derived and compared with DSWN beamformer. Indeed, when the signal-to-noise ratio is high, namely when $\text{norm}[\mathbf{A}(\hat{\Theta}^{(\hat{q})})^H\mathbf{A}(\hat{\Theta}^{(\hat{q})})] \gg \text{norm}[\hat{\sigma}^2(\hat{\mathbf{C}}^{(\hat{q})})^{-1}]$, (any sensible matrix norm will do) the above estimator reduces to

$$\hat{\mathbf{s}}(t_i) = (\mathbf{A}(\hat{\Theta}^{(\hat{q})})^H\mathbf{A}(\hat{\Theta}^{(\hat{q})})^{-1}\mathbf{A}(\hat{\Theta}^{(\hat{q})})^H\mathbf{x}(t_i), \tag{7.60}$$

which is identical in form to the optimal beamformer in the DSWN model (6.54).

It is instructive to compare this beamformer also with the Minimum Variance beamformer. To this end, observe that when the signals are uncorrelated, i.e., when $\mathbf{C} = \text{diag}\{p_1, \ldots, p_q\}$, it follows from (7.57) that the estimator of the k-th signal is given by

$$\hat{s}_k(t_i) = \hat{p}_k\mathbf{a}(\hat{\theta}_k)^H\hat{\mathbf{R}}^{-1}\mathbf{x}(t_i), \tag{7.61}$$

which is identical, up to a scalar factor, to the Minimum Variance beamformer.

However, when the signals are correlated, i.e., when the matrix **C** is non diagonal, the two estimators differ drastically. Unlike the Minimum Variance estimator which suffers severe degradation when the signals are partially correlated and fails completely when the signals are fully correlated, this estimator exploits the correlation to improve the performance.

8. SUBOPTIMAL SOLUTION TO THE SSCN MODEL

The solutions presented in the previous sections, both optimal and suboptimal, were for White Noise (WN) model. In this section we address the much more complex Colored Noise (CN) model. Specifically, we present a suboptimal solution for the Stochastic Signals (SS) and Colored Noise (CN) model.

8.1 SIMULTANEOUS DETECTION AND LOCALIZATION

As in the previous sections, our approach is based on a simultaneous solution of the detection and localization problems via the Minimum Description Length (MDL) principle for model selection. The derivation follows Wax (1989b).

A straightforward application of the MDL principle to our problem is computationally very unattractive because of the large number of unknown parameters in the SSCN model. Indeed, from (1.3) and the SS and CN model it follows that $\mathbf{x}(t)$ is a complex Gaussian random vector with zero mean and covariance matrix given by

$$\mathbf{R} = \mathbf{A}(\Theta)\mathbf{P}\mathbf{A}^H(\Theta) + \mathbf{Q}. \tag{8.1}$$

Thus, since the vector Θ is real and the matrices $\mathbf{P}$ and $\mathbf{Q}$ are Hermitian, it follows that the number of unknown real parameters in Θ, $\mathbf{P}$ and $\mathbf{Q}$, are q,

q^2 and p^2, respectively, which amount to a total of $q + q^2 + p^2$ parameters. The computation of the maximum likelihood estimator calls therefore for the solution of a $(q + q^2 + p^2)$-dimensional nonlinear maximization problem, thus rendering a straightforward application of the MDL model selection criterion very unattractive.

To circumvent this difficulty we recast the problem in the signal subspace framework and compute the description length by models based on the decomposition of $\mathbf{x}(t)$ into its components in the signal and noise subspaces. These models, though suboptimal, simplify comsiderably the computational load.

As in the optimal solutions for the WN model, we compute the description length by summing up three terms: (i) the description length of the noise subspace componenets $\{\mathbf{x}_N(t_i)\}$, assuming that $\Theta^{(k)}$ is given, (ii) the description length of the signal subspace componenets $\{\mathbf{x}_S(t_i)\}$, assuming that the noise subspace componenets $\{\mathbf{x}_N(t_i)\}$ and $\Theta^{(k)}$ are given, and (iii) the description length of $\Theta^{(k)}$.

To compute the first term we need a probabilistic model for the noise subspace components. Note that from CN model and (6.26) it follows that $\mathbf{x}_N(t)$ is a $(p - k) \times 1$ complex Gaussian random vector with zero mean. Denoting its covariance matrix by $\mathbf{R}_{NN}(\Theta^{(k)})$, we have

$$\mathbf{x}_N(t)|\Theta^{(k)} \sim N_{p-k}(\mathbf{0}, \mathbf{R}_{NN}(\Theta^{(k)})). \tag{8.2}$$

Hence,

$$f(\{\mathbf{x}_N(t_i)\}|\Theta^{(k)}) = \prod_{i=1}^{M} \frac{1}{|\pi \mathbf{R}_{NN}(\Theta^{(k)})|} \exp(-\mathbf{x}_N^H(t_i)\mathbf{R}_{NN}^{-1}(\Theta^{(k)})\mathbf{x}_N(t_i)), \tag{8.3}$$

or more compactly,

$$f(\{\mathbf{x}_N(t_i)\}|\Theta^{(k)}) = |\pi \mathbf{R}_{NN}(\Theta^{(k)})|^{-M} \exp(-M \operatorname{tr}[\mathbf{R}_{NN}^{-1}(\Theta^{(k)}) \hat{\mathbf{R}}_{NN}(\Theta^{(k)})]), \tag{8.4}$$

where $\hat{\mathbf{R}}_{NN}(\Theta^{(k)})$ denotes the $(p-k) \times (p-k)$ sample-covariance matrix of $\mathbf{x}_N(t)$,

$$\hat{\mathbf{R}}_{NN}(\Theta^{(k)}) = \frac{1}{M} \sum_{i=1}^{M} \mathbf{x}_N(t_i) \mathbf{x}_N^H(t_i). \tag{8.5}$$

As can be readily verified, the maximum likelihood estimator of $\mathbf{R}_{NN}(\Theta^{(k)})$ is given by $\hat{\mathbf{R}}_{NN}(\Theta^{(k)})$. Hence, since $\mathbf{R}_{NN}(\Theta^{(k)})$ containes $(p-k)^2$ real paprametrs, it then follows from (2.4) that the code length required to encode the noise subspace components, ignoring constant terms which are independent of k, is given by

$$L\{\mathbf{x}_N(t_i)|\Theta^{(k)}\} = M \log |\hat{\mathbf{R}}_{NN}(\Theta^{(k)})| + M(p-k) + \frac{1}{2}(p-k)^2 \log M. \tag{8.6}$$

To compute the second term, we shall construct a probabilistic model for the signal subspace components given $\Theta^{(k)}$. The dependence on the noise subspace componets will be ignored to simplify the model. To this end, observe that from the SSCN model and (6.24) it follows that, given $\Theta^{(k)}$, $\mathbf{x}_S(t)$ is a $k \times 1$ complex Gaussian random vector with zero mean. Denoting its covariance matrix by $\mathbf{R}_{SS}(\Theta^{(k)})$, we have

$$\mathbf{x}_S(t)|\Theta^{(k)} \sim N_k(\mathbf{0}, \mathbf{R}_{SS}(\Theta^{(k)})). \tag{8.7}$$

Hence our model is

$$f(\{\mathbf{x}_S(t_i)\}|\Theta^{(k)}) = \prod_{i=1}^{M} \frac{1}{|\pi \mathbf{R}_{SS}(\Theta^{(k)})|} \exp(-\mathbf{x}_S^H(t_i) \mathbf{R}_{SS}^{-1}(\Theta^{(k)}) \mathbf{x}_S(t_i)), \tag{8.8}$$

or more compactly,

$$f(\{\mathbf{x}_S(t_i)\}|\Theta^{(k)}) = |\pi \mathbf{R}_{SS}(\Theta^{(k)})|^{-M} \exp\{-M \mathrm{tr}[\mathbf{R}_{SS}^{-1}(\Theta^{(k)})\hat{\mathbf{R}}_{SS}(\Theta^{(k)})]\}, \quad (8.9)$$

where

$$\hat{\mathbf{R}}_{SS}(\Theta^{(k)}) = \frac{1}{M}\sum_{i=1}^{M} \mathbf{x}_S(t_i)\mathbf{x}_S^H(t_i). \quad (8.10)$$

The maximum likelihood estimator of $\mathbf{R}_{SS}(\Theta^{(k)})$ is given by the sample covariance matrix $\hat{\mathbf{R}}_{SS}(\Theta^{(k)})$. Thus, since the number of free parameters in $\mathbf{R}_{SS}(\Theta^{(k)})$ is k^2, it then follows from (2.16) that the description length of the signal subspace components, ignoring constant terms which are independent of k, is given by

$$L\{\mathbf{x}_S(t_i)|\Theta^{(k)}\} = M \log|\hat{\mathbf{R}}_{SS}(\Theta^{(k)})| + Mk + \frac{1}{2}k^2 \log M. \quad (8.11)$$

Combining (8.7) and (8.11) and ignoring, again, constant terms which are independent of k, the total description length of the signal and noise subspace components is given by

$$L\{\mathbf{x}_N(t_i), \mathbf{x}_S(t_i)|\Theta^{(k)}\} = M \log|\hat{\mathbf{R}}_{SS}(\Theta^{(k)})||\hat{\mathbf{R}}_{NN}(\Theta^{(k)})| + \frac{1}{2}((p-k)^2+k^2)\log M. \quad (8.12)$$

To compute the third term we have first to decide on the algorithm used for the estimation of $\Theta^{(k)}$. Indeed, since this value is unknown, it must be estimated from the data and encoded as a preamble to the code string. To derive this estimator, recall that our goal is to obtain the shortest code length, which we claim assures the best detection performance. Hence the optimal estimate of $\Theta^{(k)}$ is obtained by minimizing the description length of the data as given by (8.12),

$$\hat{\Theta}^{(k)} = \arg\min_{\Theta^{(k)}} \log(|\hat{\mathbf{R}}_{SS}(\Theta^{(k)})||\hat{\mathbf{R}}_{NN}(\Theta^{(k)})|). \quad (8.13)$$

This estimator has an interesting and intuitively appealing geometric interpretation. Observe that $|\hat{\mathbf{R}}_{SS}(\Theta^{(k)})|$ represents the volume occupied by the the data in the signal subspace while $|\hat{\mathbf{R}}_{NN}(\Theta^{(k)})|$ represents the volume occupied by the data in the noise subspace. We can therefore interprete the estimator (8.13) as the value of $\Theta^{(k)}$ that minimizes the volume occupied by the data in *both* the signal and the noise subsapces. This interpretation is intuitively pleasing since the minimization of the volume of the projections onto the noise subspace guarantees good fit of the signal subspace while the minimization of the volume of the projections onto the signal subspace guarantees the exploitation of the stochastic model of the signals, or more specifically, their being zero mean vectors.

Another form of this estimator can be obtained by casting it in terms of the eigenvalues of the matrices involved. Indeed, observe that from (7.29) and (7.30), by the well known invariance property of the determinant under the transformation $\mathbf{F} \rightarrow \mathbf{G}\mathbf{F}\mathbf{G}^H$ when $\mathbf{G}$ is unitary, we get

$$|\hat{\mathbf{R}}_{SS}(\Theta^{(k)})| = \prod_{i=1}^{k} l_i^S(\Theta^{(k)}), \tag{8.14}$$

and

$$|\hat{\mathbf{R}}_{NN}(\Theta^{(k)})| = \prod_{i=1}^{p-k} l_i^N(\Theta^{(k)}), \tag{8.15}$$

where $l_1^S(\Theta^{(k)}) \geq \ldots \geq l_k^S(\Theta^{(k)})$ denote the nonzero eigenvalues of the rank-k matrix $\mathbf{P}_{\mathbf{A}(\Theta^{(k)})}\hat{\mathbf{R}}\mathbf{P}_{\mathbf{A}(\Theta^{(k)})}$ and $l_1^N(\Theta^{(k)}) \geq \ldots \geq l_{p-k}^N(\Theta^{(k)})$ denote the nonzero eigenvalues of the rank-$(p-k)$ matrix $\mathbf{P}^{\perp}_{\mathbf{A}(\Theta^{(k)})}\hat{\mathbf{R}}\mathbf{P}^{\perp}_{\mathbf{A}(\Theta^{(k)})}$.

Hence, substituting (8.14) and (8.15) into (8.13), we get

$$\hat{\Theta}^{(k)} = \arg\min_{\Theta^{(k)}}\Big(\prod_{i=1}^{k} l_i^S(\Theta^{(k)}) \prod_{i=1}^{p-k} l_i^N(\Theta^{(k)}\Big). \tag{8.16}$$

Using this estimator in (8.12) and adding $\frac{1}{2}k \log M$ for the description length of its k real parameters, the total number of free parameters becomes $k^2 + (p - k)^2 + k = p^2 - k(2p - 2k - 1)$. Hence, dropping terms which do not depend on k, the MDL criterion is given by

$$\hat{k} = \arg\min_k MDL_{SSCN}(k), \tag{8.17}$$

where

$$MDL_{SSCN}(k) = M \log\Big(\prod_{i=1}^{k} l_i^S(\hat{\Theta}^{(k)}) \prod_{i=1}^{p-k} l_i^N(\hat{\Theta}^{(k)})\Big) - \frac{1}{2}k(2p - 2k - 1)\log M, \tag{8.18}$$

with $\hat{\Theta}^{(k)}$ denoting the MDL estimator given by (8.16).

By analogy, the AIC, Akaike (1973), is given by

$$AIC_{SSCN}(k) = M \log\Big(\prod_{i=1}^{k} l_i^S(\hat{\Theta}^{(k)}) \prod_{i=1}^{p-k} l_i^N(\hat{\Theta}^{(k)})\Big) - k(2p - 2k - 1). \tag{8.19}$$

Observe that the resulting form of the MDL criterion is rather unconventional. Firstly, the second term, referred to as the complexity term, is *negative* and is not a monotonic function of k. Rather, it has a parabolic behaviour with the peak being at $p/2$. Secondly, which is to be expected in light of the behaviour of the second term, the first term is also not a monotonic function of k. The exact behaviour of this term however is very difficult to analyze since it depends also on the data.

The computational load is relatively modest since only a k-dimensional minimization is involved and is similar to that involved in the optimal techniques for the WN model described in the previous sections.

REFERENCES

[1] Akaike, H. (1973): "Information Theory and an Extension of the Maximum Likelihood Principle," *Proc. 2nd Int. Symp. Inform. Theory*, Petrov, B. N. and Caski, F. Eds., pp. 267-281.

[2] Akaike, H. (1974): "A New Look at the Statistical Model Identification," *IEEE Trans. on AC*, Vol. 19, pp. 716-723.

[3] Anderson, T. W. (1963): "Asymptotic Theory for Principal Components Analysis," *Ann. of Math. Stat.*, Vol. 34, pp. 122-148.

[4] Bohme, J. F. (1986): "Estimation of Spectral Parameters of Correlated Signals in Wavefields," *Signal Processing*, Vol. 11, pp. 329-337.

[5] Bienvenu, J. and Kopp, L. (1979): "Principle de la Goniometric Passive Adaptive," *Proc. 7'eme Collque GRESTI*, (Nice France), pp. 106/1-106/10

[6] Bienvenu, J. and Kopp, L. (1980): "Adaptivity to Backgroundd Noise Spatial Coherence for High Resolution Passive Methods," *ICASSP 80*, (Denver, CO), pp. 307-310.

[7] Bresler, Y. and Macovski, A. (1986): "On the Number of Signals Resolvable by a Uniform Linear Array," *IEEE Trans. on ASSP*, Vol. 34, pp. 1361-1375.

[8] Borgiotti, G. V. and Kaplan, L. J. (1979): "Superresolution of Uncorrelated Interference Sources by Using Adaptive Array Techniques," *IEEE Trans. on AP*, Vol. 27, pp. 842-845.

[9] Capon, J. (1969): "High Resolution Frequency Wave Number Spectrum Analysis, *Proc. IEEE*, Vol. 57, pp. 1408-1418.

[10] Evans, J. E., Johnson, J. R. and Sun D.F. (1982): " Application of Advanced Signal Processing t Angle-of-Arrival Estimation in ATC Navigation and Survilence Systems," MIT Lincoln Lab., Lexington, MA, Rep. 582.

[11] Graham, A. (1981): *Kroneker Products and Matrix Calculus With Applications*, Elis Horwood Ltd., Chichester, UK.

[12] Hudson, J. E. (1981): *Adaptive Array Processing*, Peter Peregrinus.

[13] Hurewicz, W. and Wallman, H. (1948): *Dimension Theory*, Princeton University Press.

[14] Jaffer, A. G. (1988): "Maximum Likelihood Direction Finding of Stochastic Sources: A Separable Solution," *ICASSP 88*, pp. 2893-2296.

[15] Kaveh, M. and Barabell, A. J. (1986): "The Statistical Performance of the MUSIC and Minimum Norm Algorithms in resolving Plane Waves in Noise," *IEEE Trans. on ASSP*, Vo. 34, pp. 331-341.

[16] Kumaresan, R. and Tufts, D. W. (1983): "Estimating the Angle-of-Arrival of Multiple Plane Waves," *IEEE Trans. on AES*, Vol. 19, pp. 134-139.

[17] Monzingo, R. A. and Miller, T. W. (1980): "Introduction to Adaptive Arrays," Wiley-Interscience, New-York.

[18] Nehorai, A. Starer, D. and Stoica, P. (1990): "Consistency of Direction-of-Arrival Estimation with Multipath and Few Snapshots," *ICASSP 90*, pp. 2819-2822.

[19] Ottersten, B. and Ljung L. (1989): "Asyptotic Results for Sensor Array Processing," *ICASSP 89*, pp. 2266-2269.

[20] Porat, B. and Friedlander, B. (1988): "Analysis of the Asymptotic Relative Efficiency of the MUSIC Algorithm," *IEEE Trans. on ASSP*, Vol. 36, pp. 532-544.

[21] Reddi, S. S. (1979): "Multiple Source Location - A Digital Approach," *IEEE Trans. on AES*, Vol. 15, pp. 95-105.

[22] Rissanen, J. (1978): "Modeling by the Shortest Description," *Automatica*,

Vol. 14, pp. 465-471.

[23] Rissanen, J. (1983): "A Universal Prior for the Integers and Estimation by Minimum Description Length," *Ann. of Stat.*, Vol. 11, pp. 416-431.

[24] Rissanen, J. (1989): Stochastic Complexity in Statistical Inquiry, World Scientific, Series in Computer Science-Vol. 15.

[25] Rudin, W. (1976): Principles of Mathematical Analysis, McGraw-Hill.

[26] Schwartz, G. (1978): "Estimating the Dimension of the Model," *Ann. Stat.*, Vol. 6, pp. 461-464.

[27] Schmidt, R. O. (1979): "Multiple Emitter Location an Signal Parameter Estimation," *Proc. RADC Spectrum Estimation Workshop*, (Griffis AFB, N.Y), pp. 243-258.

[28] Schmidt, R. O. (1981): "A Signal Subspace Approach to Multiple Emitter Location and Spectral Estimation," Ph.D Dissertation, Stanford University, CA.

[29] Sharman, K., Durrani, T. S., Wax, M. and Kailath, T. (1984): "Asymptotic Performance of Eigenstructure Spectral Analysis Methods," *ICASSP 84*, pp. 45.5.1-45.5.4.

[30] Shannon, C. E. (1948): "Thw Mathematical Theory of Communication," *Bell Syst. Tech. J.*, Vol. 46, pp. 497-511.

[31] Shan, T. J., Wax, M. and Kailath, T. (1985): " On Spatial Smoothing for Direction-of-Arrival Estimation of Coherent Sources, " *IEEE Trans. on ASSP*, Vol. 33, No 4, pp. 806-811.

[32] Stoica, P. and Nehorai, A. (1989a): "MUSIC, Maximum Likelihood and the Cramer-Rao Bound," *IEEE Trans. on ASSP*, Vol. 37, pp. 720-743.

[33] Stoica, P. and Nehorai, A. (1989b): "MUSIC, Maximum Likelihood and

the Cramer-Rao Bound: Further Results and Comparisons," *ICASSP 89*, pp.

[34] Wax, M. (1985): "Detection and Estimation of Superimposed Signals," Ph.D Dissertation, Stanford University.

[35] Wax, M. (1989a): "Detection of Coherent and Noncoherent Signals via the Stochastic Signals Model," submitted to *IEEE Trans. on ASSP.*

[36] Wax, M. (1989b): "Detection and Localization of Multiple Source in Spatially Colored Noise," submitted to *IEEE Trans. on ASSP.*

[37] Wax, M. and Kailath, T. (1985): "Detection of Signals by Information Theoretic Criteria," *IEEE Trans. on ASSSP*, Vol. 33, pp. 387-392.

[38] Wax, M. and Ziskind, I. (1989a): "On Unique Localization of Multiple Sources in Passive Sensor Arrays," *IEEE Trans. on ASSP*, Vol. 37, No. 7, pp. 996-1000.

[39] Wax, M. and Ziskind, I. (1989b): "Detection of the Number of Coherent and Noncoherent Signals by the MDL Principle," *IEEE Trans. on ASSP*, Vol. 37, No. 8, pp. 1190-1196.

[40] Widrow, B., Duvall, K. M. Gooch, R. P. and Newman, W. C. (1982): "Signal Cancellation Phenomena in Adaptive Antennas: Causes and Cures," *IEEE Trans. on AP*, Vol. 30, pp. 469-478.

[41] Zhao, L. C., Krishnaiah, P. R. and Bai, Z. D. (1986): "On Detection of the Number of Signals in the Presence of White Noise," *J. Multivariate Anal.*, Vol. 20, pp. 1-20.

[42] Ziskind, I. and Wax, M. (1988): "Maximum Likelihood Localization of Multiple Sources by Alternating Projection," *IEEE Trans. on ASSP*, Vol. 36, pp. 1553-1560.

NONSTANDARD METHODS IN PREDICTION

H. V. Poor
Princeton University, Princeton, NJ, USA

ABSTRACT

Recent developments in the theory and application of signal prediction are reviewed. In particular, three recently developed general methodologies for designing predictors under nonclassical operating conditions are presented. These three methodologies are *robust prediction*, *high-speed Levinson modeling*, and *ACM nonlinear prediction.* The first of these three methodologies deals with the design of linear predictors for situations in which the spectral model for the signal to be predicted is uncertain. In this context, a general formulation is presented through which the design of effective predictors can proceed in such situations. The second methodology involves the fitting of linear stochastic models to signals that are sampled at high rates relative to their underlying dynamics. Here, a representation of discrete-time signals based on divided-difference operators is used to develop Levinson-type algorithms that remain numerically stable in this high-speed regime. The third methodology involves the application of a family of fixed and adaptive nonlinear predictors, based on the *approximate conditional mean* (ACM) technique for recursive nonlinear filtering, to signals with a non-Gaussian component. This methodology is discussed in the specific context of intereference suppression in wideband digital communication systems; and this is seen to be a much more effective approach to such problems than are the traditional fixed and adaptive linear prediction approaches heretofore applied in such applications.

OVERVIEW

The problem of devising a prediction filter, or equivalently a whitening filter, for a stochastic signal is a central problem in statistical signal processing. In addition to the utility of prediction and whitening in their own rights, this process is tantamount to producing a realization model for the signal. Thus, this is a problem that arises in a large fraction of statistical signal processing applications. The classical theory of prediction, pioneered by Wiener and Kolmogorov (see, e.g., Poor [1]), treats the design of linear prediction filters within the context of statistical models that are known and well-conditioned. Unknown models are accommodated within this classical framework by adding coefficient adaptivity to linear predictors of fixed structure (see, e.g., Goodwin and Sin [2]).

Since many applications do not fit the models for which these classical approaches are effective, signal prediction continues to be a problem of considerable research interest. In this chapter, we will provide an overview of several recent developments in the theory and application of signal prediction. In particular, three recently developed general methodologies for designing predictors under nonclassical operating conditions are presented. These three methodologies are *robust prediction*, *high-speed Levinson modeling*, and *ACM nonlinear prediction.*

Generally speaking, robust filtering deals with the design of filters under statistical criteria (such as minimum mean-squared error) for situations in which there is uncertainty in the underlying statistical model. This methodology is based on the theory of robust statistical inference (e.g., Huber [3]), and overviews of robust techniques for signal processing can be found in [4 - 6]. Robust prediction deals specifically with the design of linear predictors for situations in which the spectral model for the signal to be predicted is uncertain. In Section 1 of this chapter, we review techniques for designing effective predictors in such situations.

The second methodology considered here involves the fitting of linear recursive models to signals that are sampled at high rates relative to their underlying dynamics. This signal-processing regime is of particular interest in applications such as digital feedback control and wideband communications, in which high relative sampling rates are often dictated by system stability or format considerations rather than by signal-processing needs. Most traditional signal processing algorithms are inherently ill-conditioned when applied in such situations. Recent work of Goodwin, *et al.* [7 - 9] has shown that a representation of discrete-time signal dynamics based on divided-difference operators can lead to alternative algorithms that are numerically stable in this high-speed regime. One of the central problems that has been studied in this context is the *Levinson problem*, which is a widely used prediction-based linear stochastic modeling technique. In Section 2, we describe the application of the divided-difference-operator approach in this context.

These first two methodologies are concerned with linear prediction methods. Linear prediction filters are of considerable value due to their relative simplicity of design, implementation and analysis. Moreover, they are theoretically the ideal structures for dealing with signal and noise phenomena that can be modeled accurately as having Gaussian statistics. However, there are many phenomena arising in applications such as sonar detection, digital communications, and speech processing, that are not well modeled as being Gaussian. For such processes, nonlinear prediction can often offer substantial performance improvements over linear prediction. Unfortunately, this performance advantage typically comes at the expense of increased difficulty in design, analysis and implementation. As the third methodology to be discussed in this chapter, Section 3 describes a family of fixed and adaptive nonlinear predictors, based on Masreliez's *approximate conditional mean* (ACM) approach to nonlinear filtering [10], that overcomes some of these difficulties. Here we use the particular application of interference suppression in a wideband communications signal as a paradigm for this methodology. However, the results will also apply to other situations involving the prediction of signals with non-Gaussian components.

1 ROBUST PREDICTION

The classical theory of stationary linear prediction is based on the assumption that one has an accurate statistical model for the signal to be predicted. However, in practice one rarely has accurate statistical models of the type required for the application of this classical theory. More likely, one often has a reasonably good first-approximation to the required statistics, an approximation from which the actual statistical behavior of the phenomena of interest can deviate somewhat. Unfortunately, it has been demonstrated that apparently mild deviations from nominal statistical models can cause fairly severe degradation in the performance of prediction procedures designed around those nominal statistics. Thus, it is not always possible to design a good predictor by applying an optimum design to an approximate statistical model. Rather it is necessary to consider the uncertainty in the statistical model from the outset of the design stage in order to find procedures that perform well in the presence of such uncertainty.

A statistical procedure that performs near optimally at a nominal statistical model, and that performs acceptably for statistical models neighboring the nominal ones, is said to be *robust*. Over the past three decades there has been considerable interest in the development of methods for designing such procedures in the contexts of statistical signal processing [4 - 6] and statistical inference [3]. In this chapter we review those aspects of this methodology that are relevant in the context of the problem of robust prediction.

1.1 Classical Wiener-Kolmogorov Prediction

Consider a real, zero-mean and wide-sense stationary (w.s.s.) random signal $\{Y_k\}_{k=-\infty}^{\infty}$, representing a phenomenon for which we wish to design a predictor. We assume that this signal has a power spectral density ϕ given by

$$\phi(\omega) = \sum_{k=-\infty}^{\infty} E\{Y_0 Y_k\} e^{i\omega k}, \quad -\pi \leq \omega \leq \pi. \tag{1.1}$$

(Note that, here and in the sequel $E\{\cdot\}$ denotes expectation.)

Suppose we observe $\{Y_k\}_{k=-\infty}^{t}$ for some integer time t, and we wish to estimate the quantity Y_{t+1} by way of a linear time-invariant filtration of $\{Y_k\}_{k=-\infty}^{t}$. That is, we form an estimate

$$\hat{Y}_{t+1} = \sum_{k=-\infty}^{t} h_{t-k} Y_k, \tag{1.2}$$

where $\{h_k\}_{k=0}^{\infty}$ represents the pulse response of a causal time-invariant discrete-time linear filter. A common performance measure in such prediction is the mean squared error (MSE), which for estimates of the form (1.2) within the above assumptions is given straightforwardly by

$$MSE \equiv E\left\{\left|Y_{t+1} - \hat{Y}_{t+1}\right|^2\right\} = \frac{1}{2\pi}\int_{-\pi}^{\pi} \left|e^{i\omega} - H(\omega)\right|^2 \phi(\omega) d\omega \triangleq e(\phi; H), \tag{1.3}$$

where H is the transfer function of the prediction filter given by the Fourier series

$$H(\omega) = \sum_{k=0}^{\infty} h_k e^{i\omega k}, \quad -\pi \leq \omega \leq \pi. \tag{1.4}$$

When the power spectrum ϕ is *known*, an optimum linear predictor can be found by minimizing the MSE functional $e(\phi; H)$ of (1.3) within the causality constraint:

$$h_k = \frac{1}{2\pi}\int_{-\pi}^{\pi} H(\omega) e^{-i\omega k} d\omega = 0 \text{ for all } k < 0. \tag{1.5}$$

To do so, we assume that the spectrum of ϕ satisfies the *Paley-Wiener condition*,

$$\int_{-\pi}^{\pi} \log \phi(\omega) d\omega > -\infty. \tag{1.6}$$

This condition implies that ϕ can be factored as $\phi(\omega) = \phi^+(\omega)[\phi^+(\omega)]^*$ where $|\phi^+(\omega)|^2 = \phi(\omega)$ and where $\phi^+(\omega)$ and $\frac{1}{\phi^+(\omega)}$ are causal transfer functions (see, e.g., Ash and Gardner [11]). Using this factorization the filter $H^\dagger$ that minimizes e within the causality constraint is given by (see, Poor [1])

$$H^\dagger(\omega) = \frac{[e^{i\omega}\phi^+(\omega)]_+}{\phi^+(\omega)}, \quad -\pi \leq \omega \leq \pi, \tag{1.7}$$

where the notation $[\cdot]_+$ denotes causal truncation:

$$\left[e^{i\omega}\phi^+(\omega)\right]_+ = \sum_{k=0}^{\infty} e^{i\omega k}\frac{1}{2\pi}\int_{-\pi}^{\pi} e^{-i(k+1)\xi}\phi^+(\xi)d\xi, \quad -\pi \le \omega \le \pi. \tag{1.8}$$

Note that the filter defined in (1.8) must be causal since its Fourier series has been truncated to delete negative-index terms, and thus the cascade of this filter and $\frac{1}{\phi^+(\omega)}$ in (1.7) is causal as required. The proof of optimality of this filter can be found in [1].

The minimum value of the mean-squared prediction error is given by the well-known Kolmogorov-Szegö-Krein formula [12]:

$$e(\phi; H^\dagger) = \exp\left\{\frac{1}{2\pi}\int_{-\pi}^{\pi} \log\phi(\omega)d\omega\right\} \triangleq e^\dagger(\phi) \tag{1.9}$$

1.2 Robust Wiener-Kolmogorov Prediction

We see from (1.7) that, in order to design a MMSE linear predictor in the above model, it is necessary to know the power spectral density ϕ exactly. In practice it is usually the case that we do not know this spectrum exactly, but rather we can assume that it lies in some *class* $\mathcal{F}$ of spectral densities "neighboring" a nominal spectrum, ϕ_0. In this situation the question is not how well a predictor performs for a given spectrum, but rather it is how well the predictor performs over the entire class $\mathcal{F}$ of spectra. Since it is unlikely that the functional $e(\phi; H)$ will be constant over $\mathcal{F}$ for each filter H, this functional is no longer suitable as a single-valued performance criterion when the spectrum is not fixed. However, a natural performance criterion in this situation is the worst-case MSE functional, defined by

$$e(\mathcal{F}; H) \triangleq \sup_{\phi\in\mathcal{F}} e(\phi; H). \tag{1.10}$$

With this criterion, an optimum prediction filter for a given uncertainty class $\mathcal{F}$ thus solves the problem

$$\min_{H\in\mathcal{H}^\dagger} e(\mathcal{F}; H), \tag{1.11}$$

where $\mathcal{H}^\dagger$ denotes the set of all transfer functions satisfying the causality constraint (1.5). A predictor solving (1.11) for a given spectral class is said to be *minimax robust* over that class, and it is desirable to find such filters.

Since $e(\mathcal{F}; H)$ is defined as a maximation, the problem of (1.11) is a *minimax game*. There is a danger in using such a design criterion that the resulting solution will be too conservative since it optimizes worst-case performance. However, it has been demonstrated through numerical experimentation (e.g., Vastola and Poor [13,

14]) that for several uncertainty classes representing a wide variety of practical situations, predictors designed by this method perform near-optimally for nominal spectra and perform quite well in the worst case. These filters are thus robust in the sense discussed above. Alternatively, it has been demonstrated that predictors designed via the classical theory to be optimum for a nominal model (i.e., via (1.7)) can perform quite poorly in the face of apparently benign deviations from nominal.

The robust design problem of (1.11) has been studied in varying degrees of generality by several authors (Vastola and Poor [14], Hosoya [15], Franke and Poor [16]). Ideally, in seeking a solution to (1.11) we would like to find a *saddlepoint* solution, i.e., a solution consisting of a pair $(\phi_L; H_R^\dagger)$ with $\phi_L \in \mathcal{F}$ such that

$$e(\phi; H_R^\dagger) \leq e(\phi_L; H_R^\dagger) \leq e(\phi_L; H), \tag{1.12}$$

for all $\phi \in \mathcal{F}$ and for all $H \in \mathcal{H}^\dagger$. Note that the right-hand inequality in (1.12) implies that $H_R^\dagger$ is the *optimum* predictor (as in (1.7)) for the spectrum ϕ_L. The left-hand inequality implies that $H_R^\dagger$ achieves its worst performance over $\mathcal{F}$ at ϕ_L, and thus $e(\phi_L; H_R^\dagger)$ provides an upper bound on the performance of $H_R^\dagger$ over the uncertainty class. The existence of a saddle point is equivalent to the condition

$$\min_{H \in \mathcal{H}^\dagger} \max_{\phi \in \mathcal{F}} e(\phi; H) = \max_{\phi \in \mathcal{F}} \min_{H \in \mathcal{H}^\dagger} e(\phi; H), \tag{1.13}$$

with the validity of which the game is said to have *value*. The value of the game equals the quantity in (1.13), which is given straightforwardly by

$$e(\phi_L; H_R^\dagger). \tag{1.14}$$

(See Barbu and Precupanu [17] for a discussion of general minimax results.)

The functional $e(\phi; H)$ is linear in the spectrum ϕ and it is quadratic in the filter transfer function H. Thus, it is a *concave-convex functional*, which is a particularly nice structure for minimax problems. In particular, Ky Fan's Theorem (see Fan [18]) gives a general minimax result for such functions. This result can be exploited to give the following, which is a corollary to a result of Franke and Poor [16]:

Theorem 1.1: Suppose $\mathcal{F}$ is a convex set satisfying the following conditions.

i.) $\frac{1}{2\pi} \int_{-\pi}^{\pi} \phi(\omega) d\omega \leq w < \infty$, for all $\phi \in \mathcal{F}$,

and

ii.) there exists a spectrum $\phi \in \mathcal{F}$ satisfying $\phi(\omega) \geq \Delta$, a.e. for some $\Delta > 0$.

Then $(\phi_L; H_R^\dagger)$, with $\phi_L \in \mathcal{F}$ and $H_R^\dagger$ the optimum prediction filter from (1.7) for the spectrum ϕ_L, is a saddlepoint for (1.11) if and only if

$$e^\dagger(\phi_L) = \max_{\phi \in \mathcal{F}} e^\dagger(\phi), \tag{1.15}$$

where $e^{\dagger}$ is from (1.9).

Theorem 1.1 provides a design method for robust prediction filters. In particular, it converts the minimax problem (1.11) into the maximization problem of (1.15). Once this maximization problem is solved, the robust prediction filter $H_R^{\dagger}$ is simply the optimum predictor (1.7) for the maximizing spectrum. Since the spectrum solving (1.15) has the maximum possible minimum prediction error, it is a *least-favorable* spectrum in $\mathcal{F}$ for prediction. So the robust predictor for $\mathcal{F}$ is the optimum predictor for the least-favorable spectrum in $\mathcal{F}$. The two conditions i.) and ii.) are fairly mild since i.) simply says that the average power in the signal is uniformly bounded over $\mathcal{F}$, and ii.) requires that at least one of the possible signal spectra will be essentially bounded away from zero.

The search for a least-favorable spectrum in this context is thus reduced to the maximization of the functional

$$\frac{1}{2\pi}\int_{-\pi}^{\pi} \log \phi(\omega)d\omega \tag{1.16}$$

over $\mathcal{F}$. This quantity is the *spectral entropy* of the signal (see, e.g., Pinsker [19]), and thus the least-favorable spectrum is *maxentropic* in $\mathcal{F}$. This is intuitively reasonable since entropy is a measure of disorder, and the most disorderly signal should be the hardest to predict. Similarly, since the entropy of a process is a measure of its indeterminism, the least-favorable spectrum could be thought of as the *most indeterministic,* a term introduced by Franke [20].

The spectral entropy functional (1.16) is easily shown to be concave, and thus the maximization of (1.15) is straightforward for many constraint sets $\mathcal{F}$. An interesting example comes from consideration of the following spectral uncertainty class (see Franke [20]):

$$\mathcal{F} = \left\{\phi \left| \frac{1}{2\pi}\int_{-\pi}^{\pi} e^{-i\omega k}\phi(\omega)d\omega \right. = c_k, |k| = 0,1,\ldots,m\right\}, \tag{1.17}$$

where $c_0,\ldots,c_m$ is a set of constants. Since $\frac{1}{2\pi}\int_{-\pi}^{\pi} e^{-i\omega k}\phi(\omega)d\omega$ is the kth-lag autocorrelation ($E\{Y_l Y_{l+k}\}$) of the signal, (1.17) corresponds to the set of all spectra whose first $(m+1)$ lag autocorrelations take the given values $c_o,\ldots,c_m$. Such a model is applicable, for example, when we have measurements of a finite number of autocorrelations of the signal, as is common in practice.

To find a robust predictor for the class $\mathcal{F}$ of (1.17) we first look for the least-favorable signal spectrum by solving the constrained optimization problem

$$\max_{\phi\in\mathcal{F}} \int_{-\pi}^{\pi} \log \phi(\omega)d\omega \text{ subject to } \frac{1}{2\pi}\int_{-\pi}^{\pi} e^{-i\omega k}\phi(\omega)d\omega = c_k; \quad |k| = 0,\ldots,m. \tag{1.18}$$

(1.18) will be recognized as the well-known *maximum-entropy spectrum fitting* problem (Papoulis [21]), and its solution is straightforwardly seen to be given by

$$\phi_L(\omega) \;=\; \frac{\sigma_o^2}{|\sum_{k=0}^m a_k e^{i\omega k}|^2} \tag{1.19}$$

with $a_0 = 1$ and with $a_1, \ldots, a_m$ and σ_0^2 satisfying the Yule-Walker equations for the correlations $c_0, \ldots, c_m$ (this problem is discussed further in Section 2 of this chapter). The spectrum ϕ_L of (1.19) is the spectrum of the mth-order autoregression

$$Y_{t+1} + \sum_{k=1}^{m} a_k Y_{t+1-k} \;=\; \sigma_0 \epsilon_{t+1},\; t \in \mathcal{Z} \tag{1.20}$$

where $\{\epsilon_t\}_{t\in\mathcal{Z}}$ is a sequence of orthogonal, zero-mean and unit-variance random variables. (Here, and throughout this chapter, $\mathcal{Z}$ denotes the set of all integers.) Thus the optimum predictor for the least-favorable spectrum (1.19) is given by

$$\hat{Y}_{t+1} \;=\; -\sum_{k=1}^{m} a_k Y_{t+1-k}\;, \tag{1.21}$$

which is a *finite-length* predictor with coefficients determined from the Yule-Walker equations. That this predictor uses only m past samples is an intuitively pleasing result, since we have no knowledge of the correlation structure beyond lags of length m. (This result has an interesting analog in nonlinear prediction, as is discussed in [16]).

The above example relates the minimax formulation of robust prediction to the traditional maximum-entropy spectrum fitting problem. This is an interesting result, since the spectral uncertainty class of (1.17) arises in many applications. However, it is also of interest to consider alternative spectral uncertainty classes that reflect other, more general, uncertainty topologies. One of the most common and useful such models used in robust design is the *ε-contaminated mixture* given by

$$\mathcal{F} = \left\{ \phi \;\middle|\; \phi(\omega) = (1-\epsilon)\phi_0(\omega) + \epsilon\sigma(\omega),\ -\pi \le \omega \le \pi, \text{ and } \frac{1}{2\pi}\int_{-\pi}^{\pi} \phi(\omega)d\omega = v^2 \right\}, \tag{1.22}$$

where ϕ_0 is a known spectrum representing a nominal spectral model, σ is an arbitrary and unknown "contaminating" spectrum, ϵ is a degree of uncertainty (lying in the range 0 to 1) that is placed on the nominal model by the designer, and v^2 is the power in the signal, which is assumed to be known. This type of model allows for a fairly general type of uncertainty in a nominal spectral model, and it can be used to model uncertainty in a variety of contexts.

The robust prediction problem (1.11) has been considered by Hosoya [15] for the particular case in which $\mathcal{F}$ is of the form (1.22). However, this problem can also be solved as a special case within a more general context. In particular, a general method of finding least-favorable spectra for (1.22) and many other uncertainty sets of interest by exploiting an analogy with minimax hypothesis testing has been described in [6]. This technique is discussed in the following subsection.

1.3 Robust Prediction and Robust Hypothesis Testing

Theorem 1.1 indicates that robust predictors can be designed by designing optimum predictors for least-favorable spectra. In this section we will discuss a method of finding such least-favorable pairs through an analogy between this problem and robust hypothesis testing as developed by Huber in [22].

In order to solve the robust prediction problem, we must first maximize the spectral entropy functional (1.16) over $\mathcal{F}$. In studying this problem it is of interest to consider situations, such as that described in (1.22), in which the signal power is known but the exact shape of the spectral density is unknown. Specifically, we assume

$$\frac{1}{2\pi}\int_{-\pi}^{\pi}\phi(\omega)d\omega = v^2, \quad \text{for all } \phi \in \mathcal{F}, \tag{1.23}$$

where v^2 is the power in the signal. Such classes are of practical interest since power is an easily measured quantity, and they are of analytical interest because previously known results from robust hypothesis testing can be applied to find solutions to the robust prediction problem for spectral classes with power contraints of this type.

Because of the power constraint (1.23), the class $\mathcal{F}$ can be normalized to give an analogous class of probability densities on $[-\pi, \pi]$. In particular, we define

$$\mathcal{P} = \{p \mid p(\omega) = \phi(\omega)/2\pi v^2, \quad -\pi \le \omega \le \pi, \quad \phi \in \mathcal{F}\}, \tag{1.24}$$

and consider the following composite statistical hypothesis test concerning a random variable Ω :

$$H_0 : \Omega \sim \mathcal{U}[-\pi, \pi] \tag{1.25a}$$

versus

$$H_1 : \Omega \sim p \in \mathcal{P} \tag{1.25b}$$

where $\mathcal{U}[-\pi, \pi]$ denotes the uniform distribution on the interval $[-\pi, \pi]$. For nonparametric classes $\mathcal{P}$, minimax tests between hypotheses such as H_0 and H_1 of (1.25) are usually known as *minimax robust tests.* Typically, such a test is the optimum test for a least-favorable density $q \in \mathcal{P}$, that satisifies the condition (see, Huber [22]):

$$\int_{\{q(\omega)>\gamma\}} q(\omega)d\omega \le \int_{\{q(\omega)>\gamma\}} p(\omega)d\omega, \quad \text{for all } p \in \mathcal{P} \quad \text{and all } \gamma \ge 0. \tag{1.26}$$

(Note that such a density is *stochastically smallest* over $\mathcal{P}$.) Densities satisfying (1.26) are of interest in our context because of the following result.

Lemma 1.1 (see Poor [23]): Suppose $q \in \mathcal{P}$ satisfies (1.26). Then

$$\int_{-\pi}^{\pi} \psi(q(\omega))d\omega \geq \int_{-\pi}^{\pi} \psi(p(\omega))d\omega \tag{1.27}$$

for all $p \in \mathcal{P}$ and for all continuous concave functions ψ.

Returning to the filtering problem, we note that the spectral entropy of (1.16) can be written in the form (1.27) with $p = \phi/2\pi v^2$, and ψ the continuous concave function

$$\psi(x) = \frac{1}{2\pi}\log(2\pi v^2 x), \quad x > 0. \tag{1.28}$$

Thus, we have the following result.

Theorem 1.2 (Poor [6], Kassam [24]): Suppose that $\mathcal{F}$ satisfies the power constraints (1.23) and further that $q \in \mathcal{P}$ satisfies (1.26) for $\mathcal{P}$ defined via (1.24). Then the least-favorable spectrum in $\mathcal{F}$ for prediction is given by

$$\phi(\omega) = 2\pi v^2 q(\omega), \quad -\pi \leq \omega \leq \pi. \tag{1.29}$$

It is interesting to note that functionals of the form $\int C(p_1/p_0)p_0$ with C *convex* can be thought of as measures of distance or divergence between two probability densities p_0 and p_1 (see, e.g., [25, 26]). Thus, since the negative of every concave function is convex, Lemma 1.1 implies that a spectrum satisfying (1.26) is closest to a white (i.e, flat) spectrum in every such sense. Since a white spectrum has universal maximum entropy for a given power level (it cannot be predicted), this result has considerable intuitive appeal

On first consideration the result of Theorem 1.2 does not appear to be particularly useful since the problem of finding a pair of densities satisfying (1.26) appears to be a more difficult one than that of maximizing the spectral entropy functional. However, density pairs satisfying (1.26) are already known for many uncertainty classes of probability densities, and thus robust filtering solutions for spectral uncertainty classes that are scaled versions of these classes can be determined immediately. The following are examples of interest.

Example 1.1: ϵ-Contaminated Mixtures

Least-favorable pairs for ϵ-contaminated mixtures (1.22) of *probability densities* are given by Huber in [22] and are shown there to satisfy (1.26). These can then be translated into least-favorable spectral densities for situations in which $\mathcal{F}$ is of this form. In particular, if $\mathcal{F}$ is of the form (1.22) with parameters (ϕ_0, ϵ, v^2), then the least-favorable spectrum is given by

$$\phi_L(\omega) = \max\{(1-\epsilon)\phi_0(\omega), d'\}, \tag{1.30}$$

where d' is chosen so that ϕ_L satisfies the power constraint. Note that the least favorable spectrum here is limited from below, which results in a robust predictor that guards against modeling the signal power as being too low in any spectral region. As ϵ increases toward 1, this lower limit d' will approach the signal power v^2, and the least favorable spectrum will "melt" into a uniform or white spectrum. At this extreme, the uncertainty is so large that the signal cannot be predicted in its worst case, and the minimax criterion of (1.11) is no longer a reasonable one. (Here, adaptive techniques would be better, as described in Section 3.)

Example 1.2: Variational Neighborhoods

Another useful and intiutively pleasing model for spectral uncertainty arises from allowing the possibility of all spectra that vary from a nominal spectrum by no more than some given amount. Using the standard measure of "variational distance" this model becomes

$$\left\{\phi \;\middle|\; \frac{1}{2\pi}\int_{-\pi}^{\pi}|\phi(\omega)-\phi_0(\omega)|d\omega \le \epsilon v^2 \quad \text{and} \quad \frac{1}{2\pi}\int_{-\pi}^{\pi}\phi(\omega)d\omega = v^2\right\}, \tag{1.31}$$

where ϕ_0 and ϵ play the same roles as in the ϵ-contaminated mixture and where $v^2 = \frac{1}{2\pi}\int_{-\pi}^{\pi}\phi_0(\omega)d\omega$. This model allows for slightly more general types of uncertainty than does the ϵ-contamination model.

Using Huber's least favorables [22] for variational neighborhoods of probability densities, the least-favorable spectrum for prediction is given here via Theorem 1.2 as

$$\phi_L(\omega) = \begin{cases} c'[v^2+c']^{-1}[v^2+\phi_0(\omega)] & \text{if } \phi_0(\omega) \le c' \\ \phi_0(\omega) & \text{if } c' < \phi_0(\omega) < c'' \\ c''[v^2+c'']^{-1}[v^2+\phi_0(\omega)] & \text{if } \phi_0(\omega) \ge c'' \end{cases}, \tag{1.32}$$

where $0 \le c' \le v^2 \le c'' \le \infty$ are chosen so that ϕ_L satisfies the power constraint and so that $\frac{1}{2\pi}\int_{-\pi}^{\pi}|\phi_L - \phi_0|d\omega = \epsilon v^2$. Note that, the least favorable spectrum here is tempered from above and below, which results in a robust predictor that guards against both underestimating and overestimating the amount of signal power in the extreme spectral regions.

Example 1.3: p-Point Classes

A somewhat different type of model for spectral uncertainty is provided by the so-called *p-point* classes. A p-point class is a set of densities of the form

$$\left\{\phi \;\middle|\; \frac{1}{2\pi}\int_{\Omega_i}\phi(\omega)d\omega = p_i v^2, i = 1,2,\ldots,n\right\}, \tag{1.33}$$

where v^2 is the power constraint, $\Omega_1,\ldots,\Omega_n$ forms a partition of $[-\pi,\pi]$, and $p_1,\ldots,p_n$ are nonnegative constants satisfying $\sum_{i=1}^{n} p_i = 1$. Note that this class consists of all spectra that have a fixed amount of power in each of the spectral regions $\Omega_1, \ldots, \Omega_n$. Such a class might arise, for example, when power measurements are taken in a number of frequency bands.

Least-favorable densities for this model are discussed in [27] and [28]. These least-favorable densities are not unique, but a simple least favorable spectrum is given by the piecewise constant density

$$\phi_L(\omega) = 2\pi v^2 p_i / \int_{\Omega_i} d\omega, \text{ for } \omega \in \Omega_i, \quad i = 1, \ldots, n \,. \tag{1.34}$$

Note that this is a zonal, or piecewise constant, spectrum. Thus the corresponding least-favorable signal here can be thought of as being composed of ideally bandlimited signals occupying nonoverlapping frequency bands.

Example 1.4: Band Models

A further interesting spectral uncertainty class consists of those spectral densities (with a given amount of average power) that lie in the band bounded above and below by two known functions. Such a class can be written as

$$\left\{ \phi \,\middle|\, L(\omega) \le \phi(\omega) \le U(\omega), -\pi \le \omega \le \pi, \text{ and } \frac{1}{2\pi}\int_{-\pi}^{\pi} \phi(\omega) d\omega = v^2 \right\}, \tag{1.35}$$

where L and U are known functions and where v^2 is fixed. Note that a model such as (1.35) can be used, for example, to describe a "confidence region" around an estimated spectrum. Also note that the ϵ-contaminated model of Example 1.1 is a special case of (1.35) with $L(\omega) = (1-\epsilon)\phi_0(\omega)$ and $U(\omega) = \infty$.

Least-favorable densities for spectral band models are given by Kassam and Lim in [29]. The solution to this problem is somewhat involved and will not be repeated here. The behavior is similar to that for the ϵ-contaminated mixtures, with some variation due to the upper bound $U(\omega)$.

The above four models provide for a fairly wide variety of uncertainty types and thus are suitable for many aplications. Other models of a similar nature can be found in Kassam [30], Rieder [31], and Bednarski [32].

1.4 Conclusion

In this section we have presented an overview of the problem of robust linear prediction in the context of the Wiener-Kolmogorov problem. We have seen that this problem can be formulated as a minimax game with mean-squared error as the payoff function. This formulation leads to the design of robust prediction filters as optimum filters for least-favorable spectral models, explicit expressions for which can be found for a number of spectral uncertainty models of practical interest. Numerical investigation has shown filters designed in this way to perform quite well, in contrast to filters designed by ignoring uncertainty, whose performance can deteriorate drastically in the face of uncertainties.

The results in this section can be generalized by relaxing many of the conditions originally placed on the signal model. Examples of other problems that have been treated include robust noisy prediction (including Wiener-Komogorov and Kalman-Bucy formulations), robust multistep prediction, robust nonlinear prediction, and formulations of robustness other than the minimax one presented here. The reader is referred to [6] for an overview of these methods.

2 THE HIGH-SPEED LEVINSON PROBLEM

High-speed digital processing methods are of increasing importance in many modern systems applications. Low speed methods are usually associated with a performance penalty, but have been necessitated by technological limitations. However, recent techological advances have made these limitations less important. Unfortunately, though, most traditional signal processing algorithms are inherently ill-conditioned when applied in situations in which data are taken at sampling rates that are high relative to the dynamics of the underlying continuous-time processes being sampled. This signal-processing regime is of particular interest in applications such as digital feedback control and wideband communications, in which high relative sampling rates are often dictated by system stability or format considerations rather than by signal-processing needs. Considerable recent progress toward ameliorating such problems has been made through the use of a divided-difference operator, rather than the conventional shift operator, to represent the dynamics of sampled data. This approach leads to a novel systems calculus that allows for a unification of continuous and discrete time formulations, enables a smooth transition from sampled-data algorithms to their continuous-time counterparts, and consequently enhances the numerical conditioning of algorithms in the high-speed regime. This means of signal representation and analysis shows considerable promise relative to traditional analysis based on shift operators in the emerging era of high-speed real-time processing, and a comprehensive overview of this methodology can be found in Goodwin, *et al.* [9]. One of the central problems that has been studied in this context is the so-called *Levinson problem*, which is very closely related to the maximum-entropy spectrum fitting problem discussed in the preceding section. In this section we describe the application of the divided-difference-operator methodology in this context.

2.1 The Classical Levinson Problem

The classical Levinson autoregressive modeling problem is concerned with choosing a parameter vector $\underline{a}_n = [a_{n,0}, a_{n,1}, \ldots, a_{n,n}]^T$ with $a_{n,0} \equiv 1$ to minimize the mean-squared prediction error $E\{\epsilon_n^2(t)\}$ where $\epsilon_n(t)$ is the error in the model

$$Y_{t+1} + \sum_{k=1}^{n} a_{n,k} Y_{t+1-k} = \epsilon_n(t), \quad t \in \mathcal{Z}, \tag{2.1}$$

where $\{Y_k\}_{k=-\infty}^{\infty}$ is an observed w.s.s. signal as in the preceding section.

As noted in Section 1, the coefficients minimizing this mean-squared error are the solutions to the so-called *Yule-Walker equations*:

$$\mathbf{R}_n \underline{a}_n = \begin{pmatrix} \pi_n \\ 0 \\ \vdots \\ 0 \end{pmatrix}, \tag{2.2}$$

where $\mathbf{R}_n$ is the $(n+1) \times (n+1)$ Toeplitz matrix whose i, j^{th} element is $c_{|i-j|}$, the $|i-j|$-lag correlation coefficient of the signal. The *Levinson-Durbin algorithm* is an algorithm for recursively solving the Yule-Walker equations of successively higher order. This algorithm is given by [2]:

$$\underline{a}_{n+1} = \begin{bmatrix} \mathbf{I}_{n+1} \\ 0 \ldots 0 \end{bmatrix} \underline{a}_n - \gamma_{n+1} \begin{bmatrix} 0 \ldots 0 \\ \mathbf{J}_{n+1} \end{bmatrix} \underline{a}_n, \quad n = 0, 1, 2, \ldots, \tag{2.3a}$$

with initialization $\underline{a}_0 = 1$, where, for each positive integer k, $\mathbf{I}_k$ denotes the $k \times k$ identity matrix and $\mathbf{J}_k$ denotes the $k \times k$ matrix that has all zero entries except for 1's in its anti-diagonal. The *reflection coefficients* $\{\gamma_n\}$ are given by

$$\gamma_{n+1} = \alpha_n / \pi_n \tag{2.3b}$$

where

$$\alpha_n = [0, \ldots, 0, 1] \, \mathbf{R}_{n+1} \begin{bmatrix} \underline{a}_n \\ 0 \end{bmatrix}, \tag{2.3c}$$

and

$$\pi_n = E\{\epsilon_n^2\} = [1, 0, \ldots, 0] \, \mathbf{R}_n \, \underline{a}_n . \tag{2.3d}$$

The mean-squared error sequence $\{\pi_n\}$ satisfies the recursion

$$\pi_{n+1} = \pi_n - \alpha_n^2 / \pi_n, \quad n = 0, 1, \ldots \tag{2.3e}$$

with initialization $\pi_0 = c_0$.

The numerical stability of the Levinson-Durbin algorithm for solving (2.2) has been established by Cybenko in [33]. However, as pointed out in [33], in many cases of practical interest the matrix $\mathbf{R}_n$ is ill-conditioned, which results in unacceptable errors when the algorithm is implemented. This ill-conditioning occurs when the prediction error π_n is very small, or, equivalently, when the reflection coefficients are close to ± 1. An important case where ill-conditioning of this nature occurs is when the discrete-time signal of interest is obtained by sampling a continuous-time process at fairly rapid rates, since $\lim_{\Delta\to 0} c_{|k-l|} = c_0$ for all k and l, if the underlying continuous-time process $\{Y(t); t \in \mathcal{R}\}$ is sufficiently smooth (mean-square continuous). (Here, and in the sequel, Δ dentoes the sampling interval used to produce the discrete-time signal under study.) Since the problem is due to the poor conditioning of $\mathbf{R}_n$, it cannot be solved by using alternative algorithms, like the Schur algorithm, to solve (2.2), as noted in [33] and by Yagle and Levy in [34].

Moreover, in this sampled-data case, the following result can be proved.

Proposition 2.1 (Vijayan, *et al.* [8]) Assume that the continuous-time process satisfies the following conditions:

i.) $\{Y(t); t \in \mathcal{R}\}$ has $n-1$ mean square derivatives.

ii.) The random vector of derivatives $[Y^{(n-1)}(0), \ldots, Y^{(1)}(0),\ Y(0)]$ has a nonsingular covariance matrix.

Then,

$$\lim_{\Delta\to 0} a_{n,j} \;=\; (-1)^j \binom{n}{j}, \quad j = 0, 1, \ldots, n, \tag{2.4}$$

and

$$\lim_{\Delta\to 0} \gamma_n \;=\; (-1)^{n-1}. \tag{2.5}$$

Thus we see that, as $\Delta \to 0$, if the autocovariance function of the continuous-time process has sufficiently many derivatives, the coefficients obtained by the Levinson algorithm will converge to the binomial coefficients $(-1)^j \binom{n}{j}$ independently of the underlying process. This points to a major difficulty with the standard Levinson formulation for finite-length linear modeling; namely, the parameters of this model contain no information about the statistics of the underlying process except in terms that are of higher-order in Δ.

2.2 The High-speed Levinson Problem

In order to correct the difficulties noted above, we will consider the reformulation of the autoregressive modeling problem (2.1) in terms of the divided-difference, or *delta*, operator. To do so, we assume for the remainder of Section 2 that the signal

$\{Y_k\}_{k=-\infty}^{\infty}$ is obtained by uniformly sampling a continuous time process $\{Y(t);\ t \in \mathcal{R}\}$ at interval Δ.

A mainstay of discrete-time signal and system modeling is the represenatation of dynamics in terms of the *shift operator* q, which operates on a discrete-time sequence $\{Y_k\}_{k=-\infty}^{\infty}$ by replacing each element Y_k with

$$qY_k = Y_{k+1}. \tag{2.6}$$

The nth-order autoregressive model (2.1) can be rewritten in terms of the shift operator as

$$A_n(q)Y_{t+1-n} = \epsilon_n(t), \quad t \in \mathcal{Z}, \tag{2.7}$$

with $A_n(q) = \sum_{k=1}^{n} a_{n,k}q^{n-k}$. (Here, of course, q^ℓ denotes ℓ repeated applications of q; i.e., $q^\ell Y_k = Y_{k+\ell}$.)

For high-speed processing, it has been demonstrated (see [7 - 9]) that the numerical problems plaguing such representations can often be ameliorated by considering alternative formulations based on the use of the delta operator as the fundamental dynamical element. The delta operator, δ, for sampling interval Δ is defined as

$$\delta = \frac{q-1}{\Delta}. \tag{2.8}$$

That is, the basic dynamical step (2.6) is replaced with

$$\delta Y_k = \frac{Y_{k+1} - Y_k}{\Delta}. \tag{2.9}$$

In this context, it is of interest to consider a model of the form

$$\left(\sum_{k=0}^{n} \beta_{n,k}\delta^{n-k}\right) Y_{t+1-n} = \nu_n(t), \quad t \in \mathcal{Z} \tag{2.10}$$

where $\underline{\beta}_n = [\ \beta_{n,0}, \beta_{n,1}, \ldots, \beta_{n,n}]^T$ with $\beta_{n,0} \equiv 1$, and $\{\nu_n(t)\}_{t=-\infty}^{\infty}$ is the sequence of modeling errors in this n^{th}-order model. Note that the interation of the delta operator is

$$\delta^\ell Y_k = \frac{\delta^{\ell-1}Y_{k+1} - \delta^{\ell-1}Y_k}{\Delta}; \tag{2.11}$$

so that

$$\delta^2 Y_k = \frac{Y_{k+2} - 2Y_{k+1} + Yk}{\Delta^2}, \tag{2.12}$$

and so forth.

One motivation for considering the model (2.10) is its parallelism with the continuous time autoregressive model given by

$$dY^{(n-1)}(t)\ +\ \hat{a}_{n,1}Y^{(n-1)}(t)dt + \ldots + \hat{a}_{n,n}Y(t)dt\ =\ dW(t)$$

where $\{W(t)\ ;\ t \in \mathcal{R}\}$ is a Wiener process. Another continuous-time model that has been used in [35], [36] and [37] is the following, which is based on an integral operator:

$$dY(t) + \int_{t-T}^{t} a(T; T-(t-s))dY(s) = dW(s). \tag{2.13}$$

In [37], this model is approximated by using the standard discrete-time AR model with order $n = T/\Delta$. As $\Delta \to 0$, $n \to \infty$ and the limiting values of the discrete AR parameters $a_{n,j}$ are related to the continuous AR function $a(T;t)$. The disadvantage of this approach is that, for small Δ, the number of parameters in the model becomes very large (the number n of parameters should grow at a Δ^{-1} rate). In comparison, (2.10) gives a parsimonious parametrization that also converges to a continuous-time model.

Note that the variables $\delta^n Y_k, \delta^{n-1} Y_k, \ldots, Y_k$ are obtained by linear transformation of the variables $q^n Y_k, q^{n-1} Y_k, \ldots, Y_k$. Since $\delta^k = (q-1)^k/\Delta^k$, this transformation can be represented by $\mathbf{T}_n$, an $(n+1) \times (n+1)$ matrix whose l, k^{th} element is given by

$$(\mathbf{T}_n)_{l,k} = \frac{(-1)^{l-k}}{\Delta^{n-k}} \binom{n-k}{l-k}, \quad 0 \leq l, k \leq n. \tag{2.14}$$

(Here, we follow the convention that the binomial coefficient $\binom{n}{k} = 0$ for $k < 0$ and $k > n$.) Thus, $\mathbf{T}_n$ is an invertible lower triangular matrix whose $n \times n$ right lower submatrix is $\mathbf{T}_{n-1}$, with $\mathbf{T}_0 = 1$. The inverse of this matrix is given by

$$(\mathbf{T}_n^{-1})_{l,k} = \Delta^{n-l} \binom{n-k}{l-k}, \quad 0 \leq l, k \leq n. \tag{2.15}$$

It is straightforward to see that the vector $\underline{\beta}_n$ that solves

$$\min_{\underline{\beta}_n} E\{\nu_n^2(t)\} = ! \tag{2.16}$$

is given by

$$\underline{\beta}_n = (\mathbf{T}_n)_{0,0} \mathbf{T}_n^{-1} \underline{a}_n = \Delta^{-n} \mathbf{T}_n^{-1} \underline{a}_n, \tag{2.17}$$

where $\underline{a}_n$ solves the Yule-Walker equations (2.2).

It is the Toeplitz property of $\mathbf{R}_n$, the nth-order covariance matrix of the signal, that makes it possible to solve (2.2) recursively using $O(n^2)$ computations via (2.3). Since $\mathbf{T}_n$ is triangular, if we knew $\mathbf{R}_n$, it would be possible to solve (2.16) using $O(n^2)$ computations, by first solving (2.2) for $\underline{a}_n$ using the Levinson-Durbin algorithm, and then using (2.17) to obtain $\underline{\beta}_n$. However, in this procedure, any numerical errors in calculating $\underline{a}_n$ due to the ill-conditioning of $\mathbf{R}_n$ would carry over to the calculation of $\underline{\beta}_n$. Also, this type of calculation is not recursive in n.

Exploiting the special structure of the matrix $\mathbf{T}_n$, Vijayan, *et al.* [8] have obtained an $O(n^2)$ algorithm for solving (2.16), that only requires knowledge of the non-Toeplitz covariance matrix of $(\delta^n Y_k, \delta^{n-1} Y_k, \ldots, Y_k)$. This algorithm has the added advantage of being recursive, like the Levinson algorithm. It is summarized in the following:

Proposition 2.2 (Vijayan, *et al.* [8]) The argument solving (2.16) is given recursively (in n) by

$$\underline{\beta}_{n+1} = \mathbf{C}_n \underline{\beta}_n + \frac{1}{\Delta^2}\frac{\tilde{\gamma}_{n+1}}{\tilde{\gamma}_n}\begin{bmatrix} 0 \\ \underline{\beta}_n \end{bmatrix} - \frac{\tilde{\alpha}_n}{\tilde{\alpha}_{n-1}}\begin{bmatrix} 0\ldots 0 \\ \mathbf{C}_{n-1} \end{bmatrix}\underline{\beta}_{n-1}, \quad n = 0,1,\ldots, \tag{2.18a}$$

with initialization $\underline{\beta}_{-1} = 0$, $\underline{\beta}_0 = 1$, and $\tilde{\gamma}_0 = -\Delta^{-2}$. Here $\mathbf{C}_n$ is the $(n+2)\times(n+1)$ matrix defined by

$$\mathbf{C}_n = \frac{1}{\Delta}\begin{bmatrix} \Delta & 0 & 0 & \ldots & 0 \\ 1 & \Delta & 0 & \ldots & 0 \\ 0 & 1 & \Delta & \ldots & 0 \\ \vdots & \vdots & \vdots & \vdots & \vdots \\ 0 & 0 & \ldots & 1 & \Delta \\ 0 & 0 & \ldots & 0 & 1 \end{bmatrix}, \tag{2.18b}$$

$\tilde{\gamma}_n$ is defined by

$$\tilde{\gamma}_{n+1} = \tilde{\alpha}_n / \tilde{\pi}_n, \quad n = 1,2,\ldots \tag{2.18c}$$

with

$$\tilde{\alpha}_n = [\,0,\ldots,0,1]\ \mathbf{Q}_{n+1}\begin{bmatrix} \underline{\beta}_n \\ 0 \end{bmatrix}, \quad n = 1,2,\ldots, \tag{2.18d}$$

$$\tilde{\alpha}_0 = E\{x(t)\delta x(t)\} + E\{x^2(t)\} / \Delta, \tag{2.18e}$$

and

$$\tilde{\pi}_n = E\{\nu_n^2(t)\} = [\,1,0,\ldots,0]\ \mathbf{Q}_n \underline{\beta}_n, \quad n = 0,1,2,\ldots \tag{2.18f}$$

with $\mathbf{Q}_n$ the covariance matrix of $(\delta^n Y_k, \ldots, \delta Y_k, Y_k)$. ($\mathbf{Q}_n$ does not depend on k due to the assumed stationarity of the signal.)

We now note that (2.18) can also be written as

$$\beta_{n+1,j} = \beta_{n,j} + \frac{1}{\Delta^2}\left(\Delta + \frac{\tilde{\gamma}_{n+1}}{\tilde{\gamma}_n}\right)\beta_{n,j-1} - \frac{\tilde{\alpha}_n}{\tilde{\alpha}_{n-1}}\left(\beta_{n-1,j-1} + \frac{1}{\Delta}\beta_{n-1,j-2}\right), j = 0,\ldots,n+1, \tag{2.19}$$

where we assume that $\beta_{n,j} = 0$ for $j < 0$ and $j > n$. On comparing (2.19) with its Levinson counterpart (2.3), we see that the complexity of the new algorithm is of the same order despite the fact that the algorithm inverts a non-Toeplitz matrix.

However, unlike the Levinson coefficients, it can be shown that (under sufficient smoothness) the delta-Levinson coefficients converge to meaningful statistical parameters of the underlying process - namely, the regression coefficients of the nth mean-square derivative on those of lower orders. Moreover, in the limit (2.19) has the form

$$\beta_{n,j} = \beta_{n-1,j} + \frac{\tilde{\pi}_{n-1,0}}{\tilde{\pi}_{n-2,0}}\beta_{n-2,j-2}, \quad j = 0, \ldots, n, \tag{2.20}$$

which is the same as a Levinson type recursion for continuous-time autoregressive models derived by Pham and le Breton in [38]. Thus, the recursion takes on a meaningful (and stable) limiting form.

These favorable limiting properties suggest that the delta-Levinson formulation will be more stable numerically for small Δ than is the standard Levinson formulation. Numerical results using floating-point calculations reported in [8] support this supposition; and, in fact, the numerical performance of (2.18) for high sampling rates is considerably better than that of the classical Levinson algorithm.

This superior numerical stability of the delta model can be explained in terms of the limiting results discussed above. In particular, there is a one-to-one correspondence between the parameter vectors $\underline{a}_n$ and $\underline{\beta}_n$. However, the limiting value of $\underline{a}_n$ is independent of the statistics of the process. Hence, the useful information in the parameter vector is being "compressed". As a result of this, small perturbations in the coefficients can cause large variations in the modeling error. The delta coefficients do not suffer from this problem since their limiting values contain useful information about the process.

Thus, we see that the delta version of the Levinson problem offers a number of advantages over the standard one for high-speed processing. However, the Levinson formulation has several useful properties that are not obviously present in its delta counterpart. Such issues as lattice implementation for layered adaptivity [2], Schur realization for parallelizability [34], and direct forms for calculating reflection coefficients with covariance data [39], fall within this category of problems. Other interesting issues include estimation techniques for extracting the covariance matrix $\mathbf{Q}_n$ from noisy data (a process almost certainly to require regularization). Progress has been made on some of these issues within the delta context. For example, in [40] a parallelizable Schur-type delta algorithm has been developed by exploiting the *Toeplitz-like* structure of $\mathbf{Q}_n$, defined in the sense of Heinig and Rost [41]. This algorithm exhibits the parallelizability of the conventional Schur algorithm, while retaining the numerical advantages of the delta-Levinson algorithm.

Finally, it should be noted that alternative versions of the Levinson-Durbin algorithm have been derived, for example, by le Roux and Gueguen in [42], that are also more robust to finite precision effects. However, the delta formulation provides a formal approach to this problem, and fits it within the unified systems framework described in [9].

3 ACM NONLINEAR PREDICTION

In the preceding two sections, we have focused primarily on *linear* prediction and prediction-based modeling methods. Linear prediction filters are of considerable value due to their relative simplicity of design, implementation and analysis. Moreover, they are theoretically the ideal structures for dealing with signal and noise phenomena that can be modeled accurately as having Gaussian statistics. However, there are many phenomena arising in applications such as sonar detection, digital communications, and speech processing, that are not well modeled as being Gaussian. For such processes, nonlinear prediction can often offer substantial performance improvements over linear prediction. Unfortunately, this performance advantage typically comes at the expense of increased difficulty in design, analysis and implementation. In this section, we discuss a family of fixed and adaptive nonlinear predictors, based on the so-called *approximate conditional mean* (ACM) approach to recursive nonlinear filtering, that overcomes some of these difficulties.

Although this approach can be used in any of a number applications involving prediction in non-Gaussian situations, we will use the particular application of interference suppression in a wideband communications signal as a paradigm for this methodology. In particular, we consider the problem of predicting a discrete-time signal that is the sum of a random binary sequence, a narrowband Gaussian signal and white Gaussian noise. This problem, as originally studied in [43 - 45], was motivated by the need to suppress narrowband interference in direct-sequence spread-spectrum communication systems. However, the results will also apply to other situations where this or related probabilistic models are valid.

3.1 Recursive ACM Prediction

In a direct-sequence spread-spectrum communication system system [46], a binary data signal is modulated with a binary *pseudonoise* (PN) signal having a nearly flat spectrum before transmission; so that the transmission bandwidth is much greater than the message bandwidth. At the receiver, the incoming signal is "despread" by correlating it with the PN signal. It has been demonstrated that the narrowband interference rejection capability of spread-spectrum systems can be improved substantially by suitably processing the received signal prior to correlating it with the PN sequence. This processing is based on the following idea.

Since the spread data signal has a nearly flat spectrum, it cannot be predicted accurately from its past values unless, of course, we make use of our knowledge of the PN sequence. The interfering signal, being narrowband, can be predicted accurately.

Hence, a prediction of the received signal based on previously received values will, in effect, be an estimate of the interfering signal. By subtracting the predicted value obtained at each sampling instant from the signal received during the subsequent instant and using the resulting prediction error as the input to the correlator, the effect of the interfering signal can be reduced.

Much of the work in this area [47 - 55] has primarily involved the use of linear transversal prediction filters to suppress the narrowband interference. Such a filter forms a linear prediction of the received signal based on a fixed number of previous samples (similarly to the Levinson predictor of (2.1)). This estimate is subtracted from the delayed received signal to obtain an error signal that is used as the input to the correlator. The filter tap coefficients are typically updated using a suitable adaptive algorithm, such as the least mean squares (LMS) algorithm [2]. For the purpose of designing and analyzing such filters, the direct-sequence communication signal can be modeled as being an independent, identically distributed (i.i.d.) binary sequence. Note that such a sequence is highly non-Gaussian. Thus, as noted above, the *optimum* filter for predicting a narrowband process in the presence of such a sequence will, in general, be *nonlinear*; and the development of such a predictor provides a useful illustration of this type of problem.

For our purposes here we can model the received signal of interest (after conversion to discrete time) as

$$Y_k = s_k + i_k + n_k\ , \quad k \in \mathcal{Z}, \tag{3.1}$$

where the discrete-time sequences $\{s_k\}$, $\{i_k\}$, and $\{n_k\}$, represent the spread data signal, narrowband interference, and white Gaussian channel noise, respectively. Details of the signal structure leading to this model can be found in [44].

Under the assumption that the PN sequence behaves as if it were random, we can consider $\{s_k\}$ to be a sequence of independent and identically distributed (i.i.d.) random variables taking on values ± 1 with equal probability. Likewise, the channel noise $\{n_k\}$ can be modeled as a sequence of i.i.d. zero mean Gaussian random variables with variance σ^2. The sequences $\{s_k\}$, $\{i_k\}$, and $\{n_k\}$, can be assumed to be mutually independent.

The narrowband interference can be modeled in a number of ways. A useful model from the viewpoint of filter design is to model this process as a Gaussian autoregressive process of order p (as in (2.1)); i.e., we assume a model of the form

$$i_k \;=\; \sum_{\ell=1}^{p} \phi_\ell i_{k-\ell} \;+\; e_k\ , \tag{3.2}$$

where $\{e_k\}$ is a white Gaussian process, and where the autoregression parameters $\phi_1, \phi_2, \ldots, \phi_p$ are assumed to be constant or slowly varying.

Under this model, the received signal has a state space representation as follows.

$$x_k \;=\; \Phi x_{k-1} \;+\; w_k\ , \tag{3.3a}$$

$$Y_k = Hx_k + v_k , \tag{3.3b}$$

where

$$x_k = [\ i_k\ i_{k-1}\ \cdots\ i_{k-p+1}\ ]^T , \tag{3.3c}$$

$$\Phi = \begin{pmatrix} \phi_1 & \phi_2 & \phi_3 & \dots & \phi_{p-1} & \phi_p \\ 1 & 0 & 0 & \dots & 0 & 0 \\ 0 & 1 & 0 & \dots & 0 & 0 \\ \vdots & \vdots & \vdots & \ddots & \vdots & \vdots \\ 0 & 0 & 0 & \dots & 0 & 1 \end{pmatrix} , \tag{3.3d}$$

$$w_k = [\ e_k\ 0\ \cdots\ 0\ ]^T , \tag{3.3e}$$

$$H = [\ 1\ 0\ \cdots\ 0\ ] , \tag{3.3f}$$

and

$$v_k = s_k + n_k. \tag{3.3g}$$

If $\{v_k\}$ were a Gaussian process, then the minimum-mean-squared-error (or *conditional mean*) prediction of the received signal (and hence of the interference) would be a linear function of the received signal, and could be computed recursively via the *Kalman-Bucy* filtering equations (e.g., [1]). However, since v_k is not Gaussian but rather is the sum of two independent random variables, one of which is Gaussian and the other of which is binary, its probability density is the weighted sum of two Gaussian densities. In this case, the exact conditional mean estimator can be shown [56] to have complexity that increases exponentially (in time), which renders it unsuitable for practical implementation.

In [10], Masreliez has developed an approximate conditional mean (ACM) filter for estimating the state of a linear system with Gaussian state noise and non-Gaussian measurement noise, such as occurs in (3.3). This filter is derived by making the approximating assumption that the state *prediction density* (i.e., the conditional density of the next state given measurements up to the present) is Gaussian. Under this assumption, the optimum prediction filter predicts Y_k as $\hat{Y}_k = H\hat{x}_k$, where $\hat{x}_k$ denotes the *state* prediction in (3.3), given recursively through the update equations

$$\hat{x}_{k+1} = \Phi\bar{x}_k , \tag{3.4a}$$

$$\bar{x}_k = \hat{x}_k + M_k H^T g_k(Y_k) , \tag{3.4b}$$

where the nonlinearity g_k is given by

$$g_k(Y_k) = -\left[\frac{\partial p(\ Y_k \mid \dots, Y_{k-2}, Y_{k-1})}{\partial Y_k} \right] \cdot [\ p(\ Y_k \mid \dots, Y_{k-2}, Y_{k-1})\ ]^{-1} , \tag{3.4c}$$

with $p(Y_k \mid \ldots, Y_{k-2}, Y_{k-1})$ denoting the *measurement* prediction density (i.e., the conditional probability density of the current measurement Y_k given the past measurements $\ldots, Y_{k-2}, Y_{k-1}$); and where the matrix M_k (which is the covariance of the state prediction error) is computed via the recursion

$$M_{k+1} = \Phi P_k \Phi^T + Q_k , \qquad (3.5a)$$

$$P_k = M_k - M_k H^T G_k(Y_k) H M_k , \qquad (3.5b)$$

where the nonlinearity G_k is

$$G_k(Y_k) = \frac{\partial g_k(Y_k)}{\partial Y_k} , \qquad (3.5c)$$

and where

$$Q_k = E\{w_k w_k^T\} . \qquad (3.5d)$$

In the model of (3.1) - (3.3), the nonlinearities involved become explicitly (see [43, 44])

$$g_k(Y_k) = \frac{1}{(H M_k H^T + \sigma^2)} \left[Y_k - \hat{Y}_k - \tanh\left(\frac{Y_k - \hat{Y}_k}{H M_k H^T + \sigma^2} \right) \right] , \qquad (3.6)$$

and

$$G_k(Y_k) = \frac{1}{(H M_k H^T + \sigma^2)} \left[1 - \frac{1}{(H M_k H^T + \sigma^2)} \mathrm{sech}^2 \left(\frac{Y_k - \hat{Y}_k}{H M_k H^T + \sigma^2} \right) \right] . \qquad (3.7)$$

If one deletes the nonlinear terms in (3.6) and (3.7) (i.e., the hyperbolic tangent in (3.6) and the hyperbolic secant in (3.7)), then (3.4) - (3.5) are the Kalman-Bucy recursions. The ACM filter is thus seen to have a structure similar to that of the standard Kalman-Bucy filter. The *time updates* (3.4a) and (3.5a) are identical to those in the Kalman-Bucy filter. The *measurement updates* (3.4b) and (3.5b) involve correcting the predicted value by a nonlinear function of the prediction *residual* $Y_k - \hat{Y}_k$. This correction is essentially acting like a soft-decision feedback to suppress the digital spread-spectrum signal from the measurements. (See, [43] for an elaboration of this point.) More generally, for other non-Gaussian situations, the nature of this nonlinearity is determined by the probability density of the observation noise through (3.4c) and (3.5c).

In [43 - 45] the results of extensive computer simulations that evaluate the capability of the ACM filter to reject an autoregressive interferer are reported. When the spectrum of the interference is sharply peaked, the ACM filter is found to give significant performance gains (on the order of 10dB for a typical model) over the

Kalman-Bucy filter, although the complexity of these two filters is quite comparable. These results thus indicate that, when the statistics of the interfering signal are known to the receiver, nonlinear filtering techniques can offer considerably better interference rejection properties than do linear filters in this application.

In the above development, we have considered nonlinear filters for suppressing interfering signals whose parameters are constant and known to the receiver. In practice, the parameters of the interference are rarely known to the receiver. Also, the interference can have statistics that vary with time. Therefore, an effective suppression filter should be able to adapt itself to variations in the interference characteristics. In the following subsection, we consider adaptive nonlinear filters that can track the interfering signal and reject it, again with the idea that this problem is paradigmatic of a more general class of problems.

3.2 Adaptive ACM Prediction

In the previous subsection, it was noted that for rejecting an autoregressive (AR) interferer with known parameters, the ACM filter performs appreciably better than the optimal linear filter. Therefore, in order to suppress an interferer with unknown or time-varying parameters, an obvious method of implementing an adaptive filter would be to identify the AR parameters of the interference recursively and to carry out the ACM filtering algorithm using the estimates obtained at each instant.

For estimating the AR parameters of the interference, we note that the received signal, being the sum of an autoregressive process and a white process, is an autoregressive moving-average (ARMA) process whose AR parameters are the same as those of the interfering signal. Thus, the parameter estimation problem of interest is reduced to that of estimating the AR parameters of an ARMA process. Several algorithms for estimating these parameters recursively are discussed by Ljung and Söderström in [57]. One possible approach is the recursive maximum likelihood (RML) algorithm that has been used in [58] and [59] for adaptive signal processing problems similar to the problem of interest. However, when the model for the interfering signal has poles close to the unit circle, which is the case for narrowband signals, the parameter estimates obtained by the RML algorithm converge rather slowly. Simulations have shown the ACM filter to be sensitive to variations in the parameter values [60]. Therefore, the obvious approach of directly adapting the ACM filter using estimates of AR parameters is not promising. Hence other approaches are necessary to obtain effective adaptive nonlinear filtering algorithms. An alternative is to consider a transversal (or finite-impulse-response) version of the ACM filter.

In [50] and [54], Iltis, Li, and Milstein have applied *linear* transversal filters to the problem of narrow-band interference suppression. In these studies, the tap weights

$a_1,\ a_2,\ \ldots,\ a_n$ of an nth-order filter are adjusted using the LMS algorithm as follows:

$$\Theta_k \;=\; \Theta_{k-1} \;+\; \mu\epsilon_n(k)X_k \tag{3.8a}$$

where Θ_k is the *tap-weight vector* obtained at time k; i.e.,

$$\Theta_k \;\triangleq\; [\ a_1(k),\ a_2(k),\ \ldots,\ a_n(k)\ ]^T, \tag{3.8b}$$

X_k is the data vector,

$$X_k \;\triangleq\; [\ Y_{k-1},\ Y_{k-2},\ \ldots,\ Y_{k-n}]^T, \tag{3.8c}$$

μ is a *tuning constant* controlling the stability and rate of convergence of the algorithm, and $\epsilon_n(k)$ is the prediction error; i.e.,

$$\epsilon_n(k) \;=\; Y_k \;-\; \hat{Y}_k\ , \tag{3.8d}$$

where

$$\hat{Y}_k \;\triangleq\; X_k^T\Theta_{k-1} \tag{3.8e}$$

is the predicted value of the received signal Y_k based on the n immediate past values.

We have seen above that in the ACM filter, the predicted value of the state was obtained as a linear function of the previous estimate modified by a nonlinear function of the prediction error. We now use the same approach to modify the adaptive linear prediction filter described above. In order to show the influence of the prediction error explicitly, the equation for the linear prediction filter can be written as

$$\hat{Y}_k \;=\; \sum_{i=1}^{n} a_i(k-1)\ [\ \hat{Y}_{k-i} \;+\; \epsilon_n(k-i)\ ]. \tag{3.9}$$

We make the assumption, similar to that made in the derivation of the ACM filter, that the prediction residual $\epsilon_n(k)$ is the sum of a Gaussian random variable and a binary random variable. If the variance of the Gaussian random variable is σ_k^2, then the nonlinear transformation appearing in the ACM filter can be written as

$$\rho_k(\epsilon_n(k)) \;=\; \epsilon_n(k) \;-\; \tanh\left(\frac{\epsilon_n(k)}{\sigma_k^2}\right). \tag{3.10}$$

By transforming the prediction error using the above nonlinearity, we get a nonlinear transversal filter for the prediction of Y_k, namely,

$$\hat{Y}_k \;=\; \sum_{i=1}^{n} a_i(k-1)[\ \hat{Y}_{k-i} \;+\; \rho_{k-i}(\epsilon_n(k-i))\ ]. \tag{3.11}$$

In order to implement this prediction filter, an estimate of the parameter σ_k^2 and an algorithm for updating the tap weights must be obtained. A useful estimate for σ_k^2 is $\hat{\sigma}_k^2 = \Delta_k - 1$ where Δ_k is a sample estimate of the prediction error variance. (It should be noted that the filter can also be adapted to track unknown amplitude for the spread-spectrum signal as well.) The LMS algorithm can still be used for the tap-weight updates. Note, however, that the LMS algorithm is a gradient algorithm based on *linear* prediction. An alternative gradient algorithm was derived in [60] based on the nonlinear prediction of (3.11). However, nonlinear filters using this algorithm were found to have increased complexity and unsatisfactory performance relative to nonlinear filters using the LMS algorithm for tap-weight updates. Thus the nonlinear algorithm with LMS tap-weight updates seems to be the best overall algorithm.

Computer simulations that compare the performance of the adaptive nonlinear predictor of (3.11) with the conventional linear ones previously considered are described in [43 - 45]. These simulation results indicate that the adaptive nonlinear filter performs significantly better (on the order of 5dB for typical models) than the best adaptive linear filter with the same number of taps, and again this is accomplished without a corresponding increase in system complexity.

It is interesting to note that the predictor (3.11) can be viewed as a generalization of both linear and hard-decision-feedback (see, e.g., [62]) adaptive predictors, in which we use our knowledge of the prediction error statistics to make a soft decision about the binary signal, that is then fed back to the predictor. As noted above, the introduction of this nonlinearity improves the prediction performance over the linear version. As discussed in [43], the softening of this feedback nonlinearity improves the convergence properties of the adaptation over the use of hard-decision feedback. This, and other aspects of this problem, are discussed in [43 - 45], to which the reader is referred for further discussion.

3.3 Conclusion

In this section, we have considered the rejection of narrow-band interference in spread-spectrum systems as a paradigm to illustrate the development of nonlinear prediction filters for situations in which part of the signal to be predicted has a non-Gaussian component. In Section 3.1, for the case of an autoregressive interferer, the problem was cast as a state estimation problem with Gaussian state noise and non-Gaussian observation noise. An approximately optimum nonlinear filter based on the ACM approximation [10] was used in this case. Based on the structure of this nonlinear estimator, an adaptive nonlinear filtering algorithm was derived in Section 3.2. Computer simulations reported in [43 - 45] indicate that these predictors can be much more effective than the conventional fixed and adaptive linear filters in this application. Thus, it is anticipated that similar experience would be observed in other applications involving prediction of signals with non-Gaussian components.

REFERENCES

1. Poor, H. V.: *An Introduction to Signal Detection and Estimation,* Springer-Verlag, New York 1988.

2. Goodwin, G. C. and S. Sin: *Adaptive Filtering, Prediction and Control,* Prentice-Hall, Englewood Cliffs, NJ 1984.

3. Huber, P. J.: *Robust Statistics,* Wiley, New York 1981.

4. Kassam S. A. and H. V. Poor: Robust techniques for signal processing: A survey, *Proc. IEEE,* 73 (1985), 433-481.

5. Poor, H. V.: Robustness in Signal Detection, in: *Communications and Networks: A Survey of Recent Advances* (Eds. I. F. Blake and H. V. Poor), Springer-Verlag, New York 1986, 131-156.

6. Poor, H. V.: Robust Filtering, in: *Advances in Statistical Signal Processing - Vol. 1: Estimation* (Ed. H. V. Poor), JAI Press, Greenwich, CT 1987, 329-360.

7. Salgado, M., R. H. Middleton and G. C. Goodwin: Connection between continuous and discrete Riccati equations with applications to Kalman filtering, *IEE Proceedings-D,* 135 (1988), 28–34.

8. Vijayan, R., H. V. Poor, J. B. Moore. and G. C. Goodwin: A Levinson-type algorithm for modeling fast-sampled data, *IEEE Trans. Automat. Contr.,* 36 (1991) 314-321.

9. Goodwin, G. C., R. H. Middleton and H. V. Poor: High-speed digital signal processing and control, *Proc. IEEE,* 80 (1992), to appear.

10. Masreliez, C. J.: Approximate non-Gaussian filtering with linear state and observation relations, *IEEE Trans. Automat. Contr.,* AC-20 (1975), 107-110.

11. Ash, R. B. and M. F. Gardner: *Topics in Stochastic Processes,* Academic Press, New York 1975.

12. Snyders, J.: Error formulae for optimal linear filtering, prediction and interpolation of stationary time series, *Ann. Math. Stat.,* 43 (1972), 1935-1943.

13. Vastola, K. S. and H. V. Poor: An analysis of the effects of spectral uncertainty on Wiener filtering, *Automatica,* 28 (1983), 289-293.

14. Vastola, K. S. and H. V. Poor: Robust Wiener-Kolmogorov theory, *IEEE Trans. Inform. Theory,* IT-30 (1984), 316-327.

15. Hosoya, Y.: Robust linear extrapolations of second-order stationary processes, *Ann. Prob.,* 6 (1978), 574-584.

16. Franke, J. and H. V. Poor: Minimax Robust Filtering and Finite-Length Robust Predictors, in: *Robust and Nonlinear Time Series Analysis* (Eds. J. Franke, W. Härdle, and R. D. Martin), Springer-Verlag, Heidelberg 1984, 87-126.

17. Barbu, V. and Th. Precupanu: *Convexity and Optimization in Banach Spaces,* Editura Academici, Bucharest 1978.

18. Fan, K.: Minimax theorems, *Proc. Nat. Acad. Sci.,* 39 (1953), 42-47.

19. Pinsker, M. S.: *Information and Information Stability of Random Variables and Processes,* Holden-Day, San Francisco 1964.

20. Franke, J.: Minimax robust prediction of discrete time series, *Z. Wahr. verw. Geb.,* 68 (1985), 337-364.

21. Papoulis, A.: Maximum entropy and spectrum estimation: A review, *IEEE Trans. Acoust. Speech and Signal Processing,* ASSP-29 (1981), 1176-1186.

22. Huber, P. J.: A robust version of the probability ratio test, *Ann. Math. Stat.,* 36 (1965), 1753-1758.

23. Poor, H. V.: On robust Wiener filtering, *IEEE Trans. Automat. Contr.,* AC-25 (1980), 531-536.

24. Kassam, S. A.: Robust hypothesis testing and robust time series interpolation and regression, *J. Time Series Analysis,* 3 (1982), 185-194.

25. Ali, S. M. and S. D. Silvey: A general class of coefficients of divergence of one distribution from another, *J. Roy. Stat. Soc., Series B,* 28 (1966), 131-142.

26. Csiszar, I.: Information-type measures of difference of probability distributions and indirect observations," *Studia Scientarium Mathematicerium Hungarica,* 2 (1967), 229-318.

27. Vastola, K. S. and H. V. Poor: On the *p*-point uncertainty class, *IEEE Trans. Inform. Theory,* IT-30 (1984), 374-376.

28. Cimini, L. J. and S. A. Kassam: Robust and quantized Wiener filters for *p*-point spectral classes, in: *Proc. 1980 Conf. Inform. Sciences and Systems,* Princeton University, Princeton, NJ 1980, 314-318.

29. Kassam, S. A. and T. L. Lim: Robust Wiener filters, *J. Franklin Inst.,* 304 (1977),

171-185.

30. Kassam, S. A.: The bounded p-point classes in robust hypothesis testing and filtering, in: *Proc. 20th Annual Allerton Conf. on Comm., Control and Computing,* University of Illinois, Urbana, IL 1982, 526-534.

31. Rieder, H.: Least favorable pairs for special capacities, *Ann. Statist.,* 5 (1977), 909-921.

32. Bednarski, T.: On solutions of minimax test problems for special capacities," *Z. Wahr. verw. Geb.,* 58 (1981), 397-405.

33. Cybenko, G.: The numerical stability of the Levinson-Durbin algorithm for Toeplitz systems of equations, *SIAM J. Scient. Statist. Comp.,* 1 (1980), 303-319.

34. Yagle, A. E. and B. C. Levy: The Schur algorithm and its applications, *Acta Applicandae Mathematicae,* 3 (1985), 255–284.

35. Kailath, T., A. Vieira, and M. Morf: Inverses of Toeplitz operators, innovations, and orthogonal polynomials, *SIAM Rev.,* 20 (1978), 106-119.

36. Kailath, T., B. C. Levy, L. Ljung, and M. Morf: Fast time-invariant implementations of Gaussian signal detectors, *IEEE Trans. Inform. Theory,* IT-24 (1978), 469-477.

37. Dewilde, P., A. C. Vieira, and T. Kailath: On ageneralized Szegö-Levinson realization algorithm for optimal linear predictors based on a network synthesis approach, *IEEE Trans. Circuits Syst.*, CAS-25 (1978), 663-675.

38. Pham, D. T. and A. le Breton: Levinson Durbin type algorithms for continuous time autoregressive models and applications, *Mathematics of Control, Signals, and Systems,* 4 (1991), 69-79.

39. Rialan, C. P., and L. L. Scharf: Fixed-point error analysis of the lattice and the Schur algorithms for the autocorrelation method of linear prediction, *IEEE Trans. Acoust., Speech, and Signal Processing,* 37 (1989), 1950-1957.

40. Vijayan, R. and H. V. Poor: Fast triangular factorization of covariance matrices of differenced time series, *SIAM J. Matrix Anal. Appl.,* to appear.

41. Heinig, G. and K. Rost: *Algebraic Methods for Toeplitz-like Matrices and Operators,* Springer-Verlag, Berlin 1988.

42. le Roux, J. and C. Gueguen: A fixed point computation of partial correlation coefficients, *IEEE Trans. Acoust., Speech, and Signal Processing,* ASSP-25 (1977), 257-259.

43. Poor, H. V. and R. Vijayan: Analysis of a Class of Adaptive Nonlinear Predictors, in: *Advances in Communications and Signal Processing* (Eds. W. A. Porter and S. C. Zak), Springer-Verlag, New York 1989, 231-241.

44. Vijayan, R. and H. V. Poor: Nonlinear techniques for interference suppression in spread-spectrum systems, *IEEE Trans. Commun.*, 38 (1990), 1060-1065.

45. Garth, L. M. and H. V. Poor: Narrowband interference suppression in impulsive channels, *IEEE Trans. Aerosp. Electron. Syst.*, 28 (1982), to appear.

46. Cooper, G. R. and C. D. McGillem: *Modern Communications and Spread Spectrum,* McGraw-Hill, New York 1986.

47. Hsu, F. M. and A. A. Giordano: Digital whitening techniques for improving spread-spectrum communications performance in the presence of narrow-band jamming and interference, *IEEE Trans. Commun.*, COM-26 (1978), 209-216.

48. Ketchum J. W. and J. G. Proakis: Adaptive algorithms for estimating and suppressing narrow-band interference in PN spread-spectrum systems, *IEEE Trans. Commun.*, COM-30 (1982), 913-924.

49. Li, L-M. and L. B. Milstein: Rejection of narrow-band interference in PN spread-spectrum systems using transversal filters, *IEEE Trans. Commun.*, COM-30 (1982), 925-928.

50. Li, L-M. and L. B. Milstein: Rejection of pulsed CW interference in PN spread-spectrum systems using complex adaptive filters, *IEEE Trans. Commun.*, COM-31 (1983), 10-20.

51. Masry, E.: Closed-form analytical results for the rejection of narrow-band interference in PN spread-spectrum systems - Part I: Linear prediction filters, *IEEE Trans. Commun.*, COM-32 (1984), 888-896.

52. Iltis, R. A. and L. B. Milstein: Performance analysis of narrow-band interference rejection techniques in DS spread-spectrum systems, *IEEE Trans. Commun.*, COM-32 (1984), 1169-1177.

53. Masry, E.: Closed-form analytical results for the rejection of narrow-band interference in PN spread-spectrum systems - Part II: Linear interpolation filters, *IEEE Trans. Commun.*, COM-33 (1985), 10-19.

54. Iltis R. A. and L. B. Milstein: An approximate statistical analysis of the Widrow LMS algorithm with application to narrow-band interference rejection, *IEEE Trans. Commun.*, COM-33 (1985), 121-130.

55. Masry E. and L. B. Milstein: Performance of DS spread-spectrum receiver em-

ploying interference-suppression filters under a worst-case jamming condition, *IEEE Trans. Commun.,* COM-34 (1986),13-21.

56. Sorenson, H. W. and D. L. Alspach: Recursive Bayesian estimation using Gaussian sums, *Automatica,* 7 (1971), 465-479.

57. Ljung, L. and T. Söderström: *Theory and Practice of Recursive Identification,* MIT Press, Cambridge, MA 1983.

58. Friedlander, B.: System identification techniques for adaptive signal processing, *IEEE Trans. Acoust., Speech, and Signal Processing,* vol. ASSP-30 (1982), 240-246.

59. Friedlander, B.: A recursive maximum likelihood algorithm for ARMA line enhancement, *IEEE Trans. Acoust., Speech and Signal Processing,* ASSP-30 (1982), 651-657.

60. Vijayan, R. and H. V. Poor: Improved algorithms for the rejection of narrowband interferers from direct-sequence signals," in: *Proc. 22nd Annual Conference on Information Sciences and Systems,* Princeton University, Princeton, NJ 1988, 851-856.

61. Li, L-M. and L. B. Milstein: Rejection of CW interference in QPSK systems using decision-feedback filters," *IEEE Trans. Commun.,* COM-31 (1983), 473-483.

